THE
ELEMENTS
OF
Mechanical Design

BY
JAMES G. SKAKOON

ASME Press New York 2008

Library of Congress Cataloging-in-Publication Data

Skakoon, James G.
The elements of mechanical design / by James G. Skakoon.
p. cm.
Includes bibliographical references and index.
ISBN 978-0-7918-0267-0
1. Machine design. 2. Engineering design. I. Title.

TJ230.S546 2008
621.8'15--dc22

2007040811

TABLE OF CONTENTS

PART III
SOME PRACTICAL ADVICE

PREFACE

This book contains principles and practices for mechanical designers. They come from the experience, know-how, and intuition of expert designers, but they represent engineering fundamentals in a practical way.

Consider two examples. Even children quickly learn that carrying two pails of water, one on each side, is easier than carrying a single pail on the right or left. This is not an isolated observation, but a useful principle of design described in this book (self-help). Or haven't we all spilled a drinking glass on a table that teeters side-to-side on two legs, resting now on the third, now on the fourth? Four-legged tables are fundamentally flawed and represent another design principle described in this book (over-constraint). Agreed, these cases are obvious enough. But in this book they become principles and guidelines, explicitly stated, to be applied to other design problems—where they may be rather less obvious.

This book is not about engineering science. Established books exist for machine design, structural analysis, and kinematics. Neither is it about design for manufacturability nor about the design process; excellent books have been written, especially in recent years, for these subjects as well. Nonetheless, despite an increased emphasis on design and manufacturing in both university curricula and practical literature, existing books have little about mechanical design as practiced by experienced designers.

Good designers often understand and use the ideas in this book whether or not they recognize them as distinct principles. Many designers, myself included, learned them either by trial and error or by exploiting colleagues' experience. Perhaps everything here would be part of all designers' practice were they to design long enough. But all too often designers are unaware of or do not fully grasp ideas that experts use to great advantage.

Therefore, in this book I explicitly state design principles and practices: 1) so beginning designers do not have to discover them on their own as I had to, and 2) so all designers can apply them as fundamental concepts throughout their designs.

Although nothing in this book is new, its narrow focus on basic, detail-level mechanical design is unique. Use it as a primer, and a refresher, on good mechanical design.

James G. Skakoon

ACKNOWLEDGEMENTS

I would like to acknowledge the people who generously agreed to be interviewed for this book:

Benge Ambrogi	Larry Gray
Joel Bartholomew	Michael Knipfer
Douglass Blanding	Herbert Loeffler
David Chastain	Brian Nelson
Ron Franklin	David Schuelke
Gilbert Fryklund	

I interviewed designers to confirm and add to the book's content. Comments repeated themselves with remarkable consistency, but many were unique. You can read some of the good ones in Appendix D.

Also generously, Paul D. Lucas, Herbert Loeffler, and Gilbert Fryklund reviewed the manuscript, as did many others anonymously, for which I am grateful.

PART I

ELEMENTARY RULES OF MECHANICAL DESIGN

1. Create designs that are explicitly simple— keep complexity intrinsic.

You have heard this often, usually as "Keep it simple!" But for good designers, just keeping it simple is not enough. If you only just keep things simple, you will still have complicated designs. You must simplify, simplify, simplify!

So what makes a design simple? Can your intuition alone judge simplicity? Will you know it when you see it?

The less thought and the less knowledge a device requires, the simpler it is. This applies equally to its production, testing, and use. Use these criteria—how much thought, how much knowledge—to judge your designs. Judge best by comparing one solution to another. Of course, it may take lots of thought and knowledge to get to a design requiring little of either; that is design.

But some devices are elaborate affairs, and you may have to design one. How can it possibly be simple amid great complexity?

What a simple design means is that everyone involved with its production and use sees nothing that looks complicated from his or her own perspective or convention. Complexity is buried and invisible. In other words, there is a hierarchy for knowledge and thought. Each hierarchical level may be intrinsically complex, yet the device remains simple if the complexity resides only within its own level.

Screw threads are a perfect example. Despite their abundant scientific and manufacturing complexity, you and I specify screw threads and threaded fasteners with the click of a mouse. Whatever it takes to make them is largely invisible to designers. We just say, "¼-20."

Another, very contemporary example is part geometry. Complex geometry no longer implies a complicated design. Computerized methods can control tooling and manufacturing processes without human interpretation. Yes, the part's geometry is complicated, yet its production and use may be no more difficult than for one with much simpler geometry. The designer easily understands the geometry; no one else needs to. Complex geometry has a cost, of course, but the hierarchical nature of knowledge permits complexity within a level if for some gain.

Simplicity can be subtle. A good example of keeping things simple is designing symmetry into components. If a part is asymmetric, knowledge and thought are needed to orient it (Figure 1-1). This comes from the assembler or user during hand assembly, or by part orientation and feed equipment during automated assembly. Assembling or using symmetrical parts requires less knowledge and thought than asymmetric parts.

Double-sided keys
insert two ways

Figure 1-1 Symmetric items are simpler—to use or assemble,
even if they are more complicated to produce.

Simplicity can be paradoxical. Symmetry adds information to the component part, thus adding cost; the double-sided key may cost more. So although the symmetrical key is simpler for the user, it is more complicated to manufacture, a different hierarchical level. But manufactured assemblies usually favor assembly simplicity over component simplicity.

Two common techniques for keeping things simple deserve specific mention:

1. Purchasing rather than making components
2. Specifying components by standards

These are good techniques not because they reduce the overall complexity, but because they hide some of it. Substantial information exists, but resides implicitly in components, most of which is no concern to designers.

2. Keep the functions of a design independent from one another.

During the concept development stage of the design process, you will decompose a device or system into basic functions [3, 4]. (If you don't, you should!) After that, a vital part of your mission as a mechanical designer is to keep those functions separate [5, 6].

This is not as easy as it sounds. There will be misunderstandings that unsuitably combine functions and compromise the design. There will be temptations to combine functions and features, resulting in a confusing overlap. Avoid the misunderstandings and resist the temptations.

Figure 2-1 shows a ball-and-socket style locking locator as might be used to hold, for example, vise jaws, dial indicators, or, in this case, a camera.

Two apparent functions are:

1. Define the camera position
2. Lock the camera in place

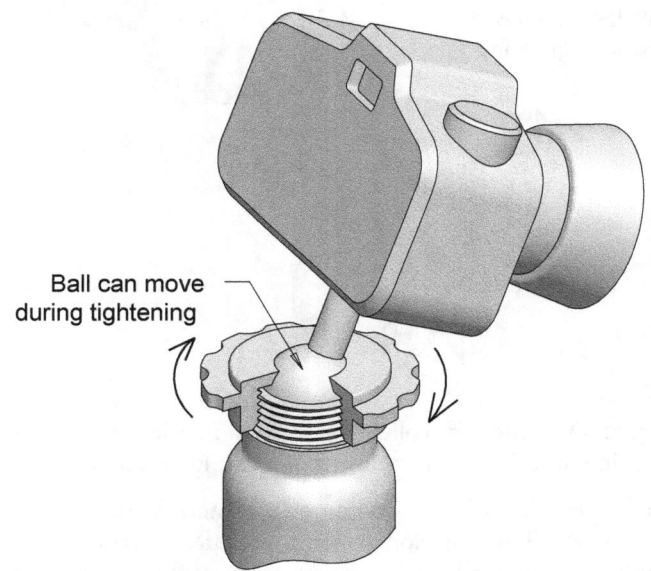

Ball can move during tightening

Figure 2-1 Ball-and-socket tripod head for camera. There is no functional independence of positioning and locking.

Although this embodiment is simple—it has only three easily manufactured parts—the above functions are not independent. Tightening the collar moves the ball in the socket because of friction between the nut and the ball. There is a race between the ball-to-socket friction and that of the nut-to-ball. Lose this frictional race, and the ball moves as the nut is tightened—quite perplexing for finely-positioned devices. You get it exactly where you want it, then you find it moved when you locked it down!

Figure 2-2 shows an improvement in separating functions. In this embodiment, tightening the nut creates no rotational friction between nut and ball, which eliminates one dependence. But all things have their cost! This design requires an over-center, or undercut geometry as well as relieving slots for the socket—costly features.

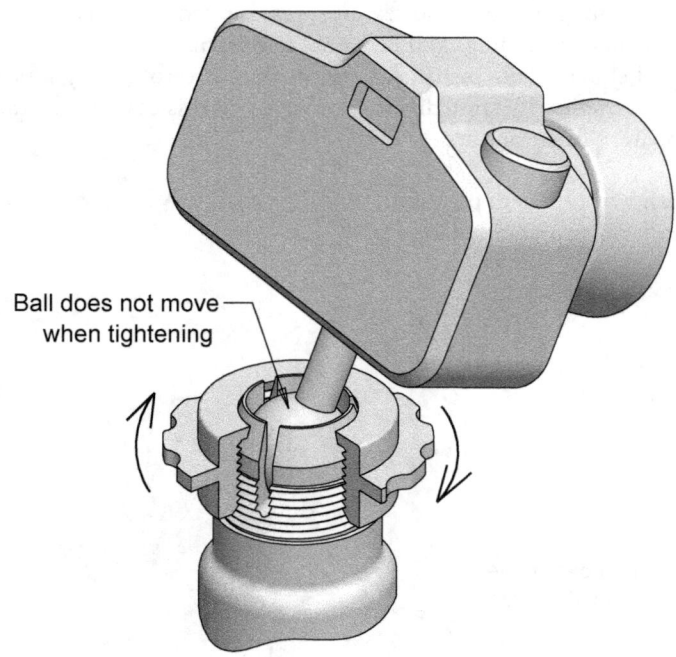

Ball does not move
when tightening

Figure 2-2 Slotted collet for ball-and-socket tripod head.
Locking function is independent of locating function.

But a deeper look suggests decomposing functions differently. The ball-and-socket joint allows rotation about three different (and non-stationary) axes simultaneously: pitch, roll, and yaw. Perhaps these should be independent?

"But this is a great design!" you say. "I have infinite possibilities for adjustment, and I can position the camera—all in one motion—right where I want it!" Yes, but you are forced to adjust pitch, roll, and yaw every time you adjust any one of them. No functional independence there!

A different solution results by decomposing functions as follows:

1. Position and fix pitch
2. Position and fix roll
3. Position and fix yaw

An example of one common approach to this set of functional requirements is a multi-axis pan head for camera tripods, shown in Figure 2-3.

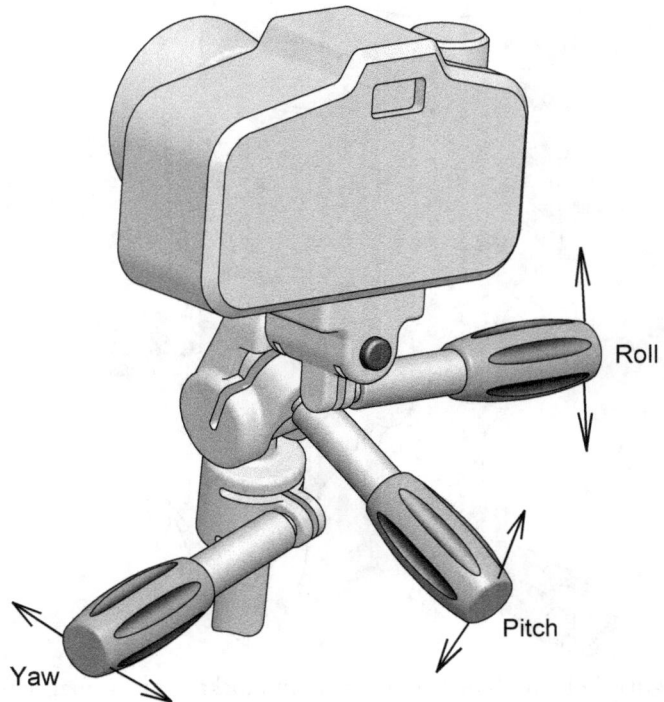

Figure 2-3 Multi-axis pan head camera mount has functional independence of all three rotational locking and positioning functions.

The nice thing here is that you adjust only what needs it, which is less awkward for fine adjust, if sometimes slower, than a ball-and-socket. The pitch, roll, and yaw adjustments are functionally independent.

If you are familiar with these examples, you already know that camera tripods with both ball-and-socket heads and multi-axis heads are successfully marketed. Users adapt to the functional dependence, and some, apparently, prefer flexibility over a more theoretically pure solution. Perhaps each is superior to the other for a specific use and specific user.

Accept that everything in design, including independence of function, is a compromise. You need not take an unyielding posture to this or any other rule of design, but always understand how and why you yield.

Finally, seek independence of the functions of a device, but do not preclude combining functions within parts. The locking handles of the multi-axis pan head offer a perfect example of this. These handles are used for two functions, positioning and locking (Figure 2-4).

Combining multiple functions into single parts saves cost and reduces complexity. This makes the cited example quite elegant and intuitive.

Figure 2-4 Combined adjustment and locking functionality reduces parts count and operates intuitively.

---

3.   Use exact constraint when designing structures and mechanisms—never overconstrain a design.

3.1   Exact constraint: a description

"Never start a fight," is good advice for life—and for mechanical design [7]. It is the essence of exact constraint, also known variously as kinematic design [7] and minimum constraint design [8].

If you rigidly constrain a component at more places than are needed, you will start a fight between these places. This is over-constraint. More scientifically, exactly constrained designs are statically determinate, whereas overconstrained or underconstrained designs are statically indeterminate [8, 10]. Exact constraint means applying just enough constraints to define a position or motion—no more, no less.

I first discovered over-constraint as a very young carpenter. Using a wood screw to fasten two wooden boards, I could not tighten the boards together, at least not at the 90° angle I wanted. The boards were either square and loose, or tight and angled, but never perfectly square and tight (Figure 3-1).

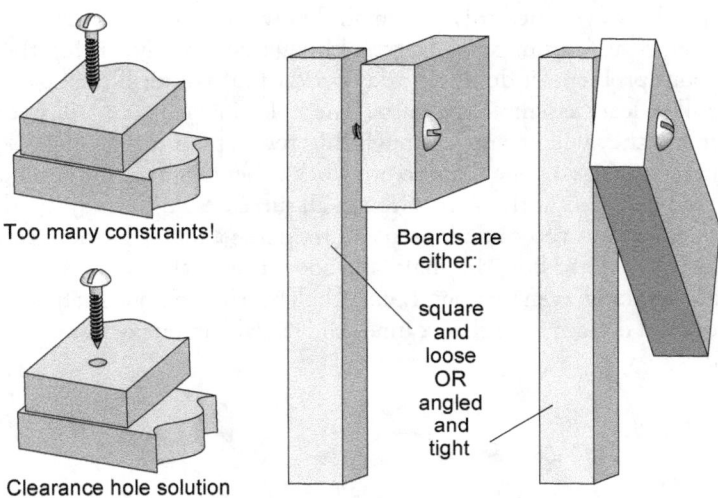

Figure 3-1 An overconstrained design.

A clearance hole, which I had not used, is the common solution, but a relieved, round shank screw works too. Either removes the axial positioning constraint in the first board so it can tighten to the second without turning with the screw.

It is tempting when attaching a hub to a shaft to pre-drill both and use a dowel or spring pin (Figure 3-2).

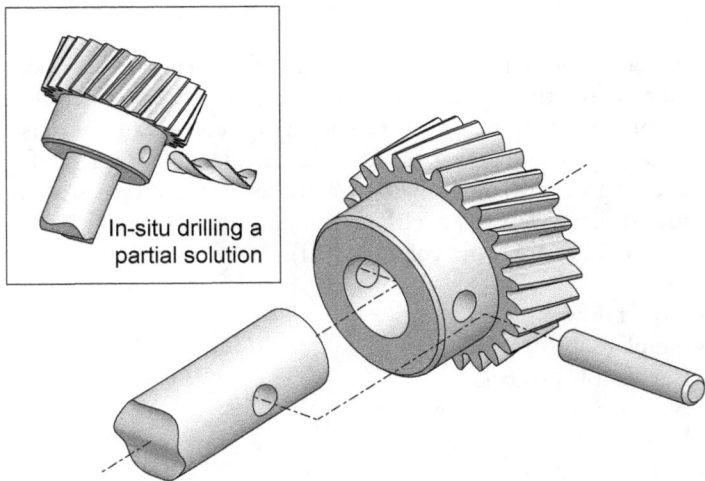

Figure 3-2 Overconstrained hub and shaft with pre-drilled holes. The holes can never line up perfectly.

If you have ever tried this, you might know it doesn't work. The holes do not line up, and drilling one larger, although apparently solving the over-constraint problem (it doesn't!), adds backlash. If you drill the hole in situ, you will at least assemble the parts. The holes in each part will not be on center, but they will line up, although this, too, is poor design.

Three bearings on one shaft do not work either (Figure 3-3). It is not luck you need trying to fit the shaft through all three bearings, it's sympathy—it won't go. Perhaps you could loosen and retighten a pillow block to assemble the shaft, but you'd bow the shaft, and sooner or later wreck a bearing too. Strictly speaking, even two rigidly-fixed ball bearings on one shaft is an over-constrained design if the shaft cannot slide axially in one or the other inner races.

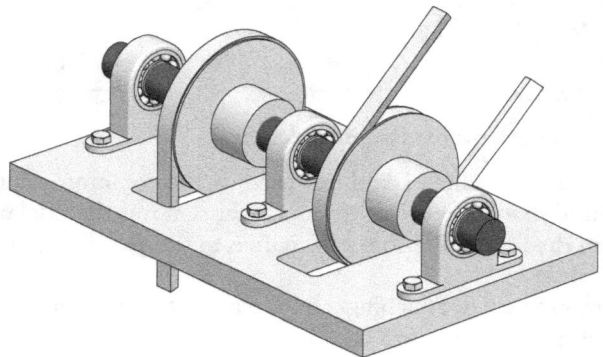

Figure 3-3 Three bearings on one shaft: an overconstrained design.

Inexperienced designers often overkill like this, bringing over-constraint into the mess, then try to patch over the symptoms. They tighten tolerances or add assembly adjustments, all to shoddy outcomes. In this case, good designers would use only two bearings with an adequately stiff shaft.

You accrue great advantages with exactly constrained compared to over-constrained designs (reprinted with permission from *Designing Cost Efficient Mechanisms* ©1993 SAE International [8]):

- ◆ No binding
- ◆ No play
- ◆ Repeatable position
- ◆ No internal stress (from assembly)
- ◆ Loose-tolerance parts
- ◆ Easy assembly
- ◆ Robustness to wear and environment

3.2 Basic theory of exact constraint

A three-dimensional object has six degrees of freedom: three translations and three rotations. These are called X, Y, Z, Θ_x, Θ_y, and Θ_z for the three translation and three rotation directions, respectively (Figure 3-4). Selectively constraint these degrees of freedom to obtain the desired motion or structure.

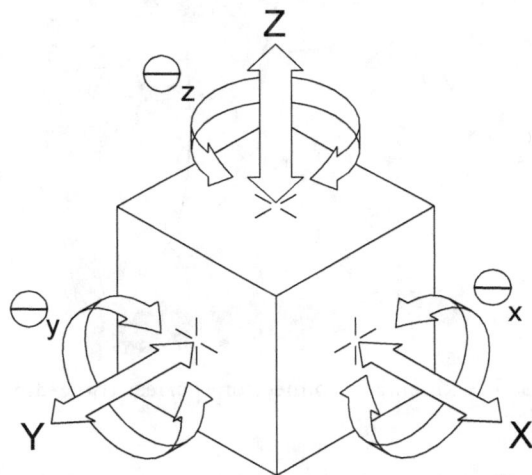

Figure 3-4 The six degrees of freedom. (Reprinted with permission from Eastman Kodak Company [9].)

A common example for illustrating exact constraint is a kinematic connection, which clearly shows all six constraints (Figure 3-5).

The first three constraints come from sphere 1 contacting three surfaces in a trihedral receptacle. These are the three translation constraints. The V-shaped groove with sphere 2 supplies two rotational constraints, and the plate surface with sphere 3, the final. You also need a nesting force, the plate's weight, for example. (There is more on nesting forces later, because it is not always that simple!) Remove the plate and it returns to the base in exactly the same position. No precision dimensions are required: ball diameter, ball position, socket positions, trihedral and V-groove dimensions and angles can vary widely without compromising exact constraint.

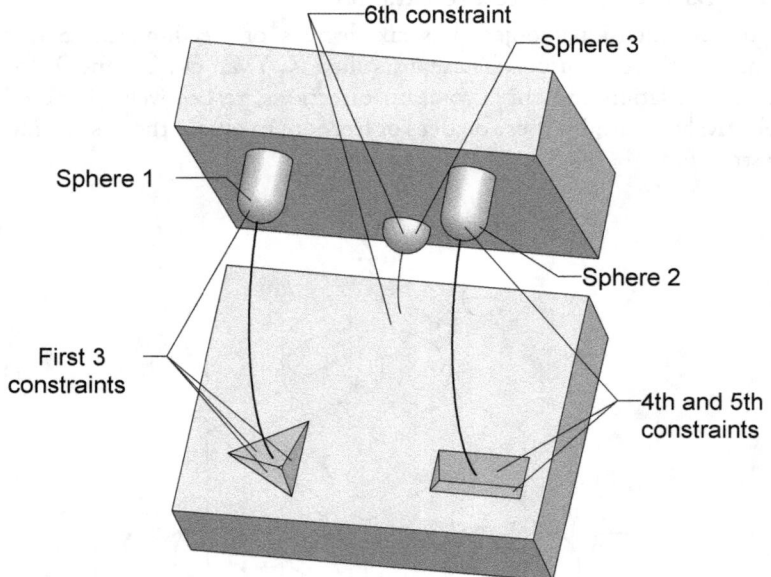

Figure 3-5 A kinematic connection: perfect three-dimensional constraint.

3.3   Exact constraint in two dimensions

Exact constraint is easier to picture in two dimensions than in three. The principles are the same, but in two-dimensions there are three degrees of freedom: two translation and one rotation (X, Y, and Θ_z).

First, we need a definition: A constraint is a point of contact maintained by a nesting force wherein the nesting force vector goes through the contact point normal to the surface of contact (Figure 3-6). Not so complicated; it is probably just what you expected.

With one constraint, the plate can still translate in Y and rotate about Z. Adding a second post can either prevent rotation (Figure 3-7) or prevent translation (Figure 3-8).

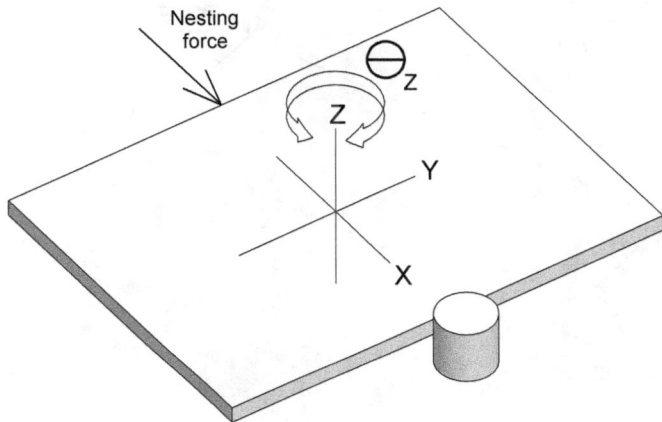

Figure 3-6 Two-dimensional example: a single constraint. A constraint is a point of contact together with a nesting force. The nesting force goes through the contact point in the tangent normal direction.

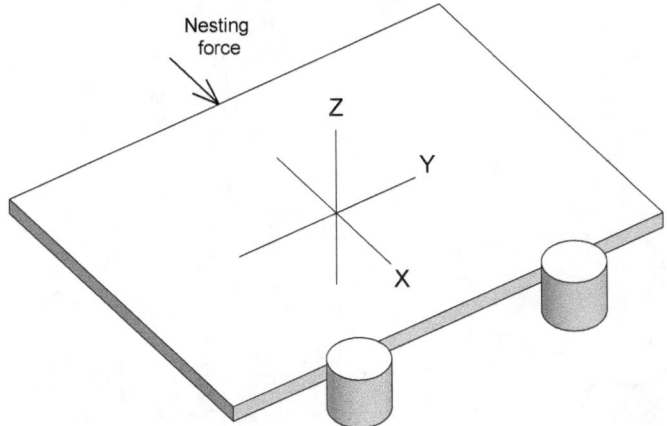

Figure 3-7 Plate constrained against rotation in two dimensions.

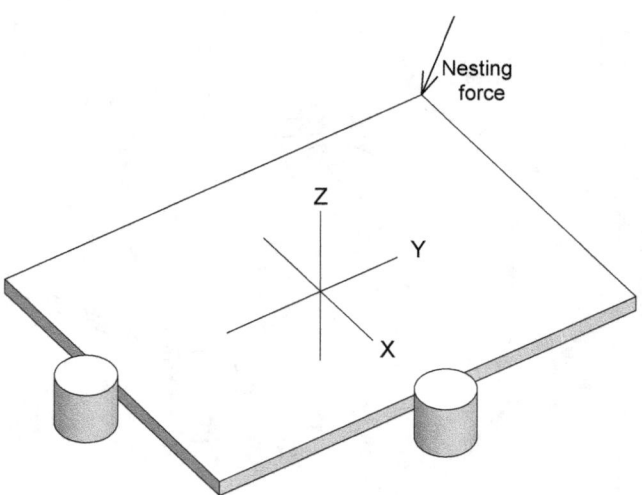

Figure 3-8 Plate constrained against translation in two dimensions.

Add a third post, and the plate has a single, unambiguous position in 2D space (Figure 3-9). That is, as long as the nesting force is enough to resist any applied forces. (Remember, more on nesting forces later!)

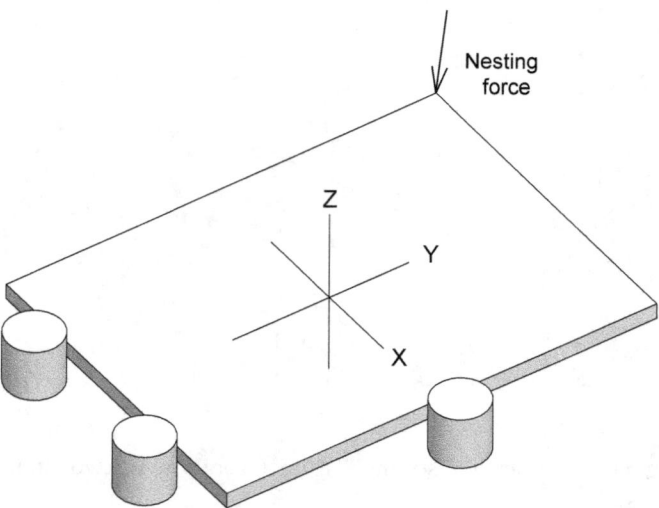

Figure 3-9 Plate fully constrained in two dimensions.

You might test your eye at judging over-, under-, and exact constraint (Figure 3-10). Appendix A lists two-dimensional and three-dimensional geometry rules for exact constraint.

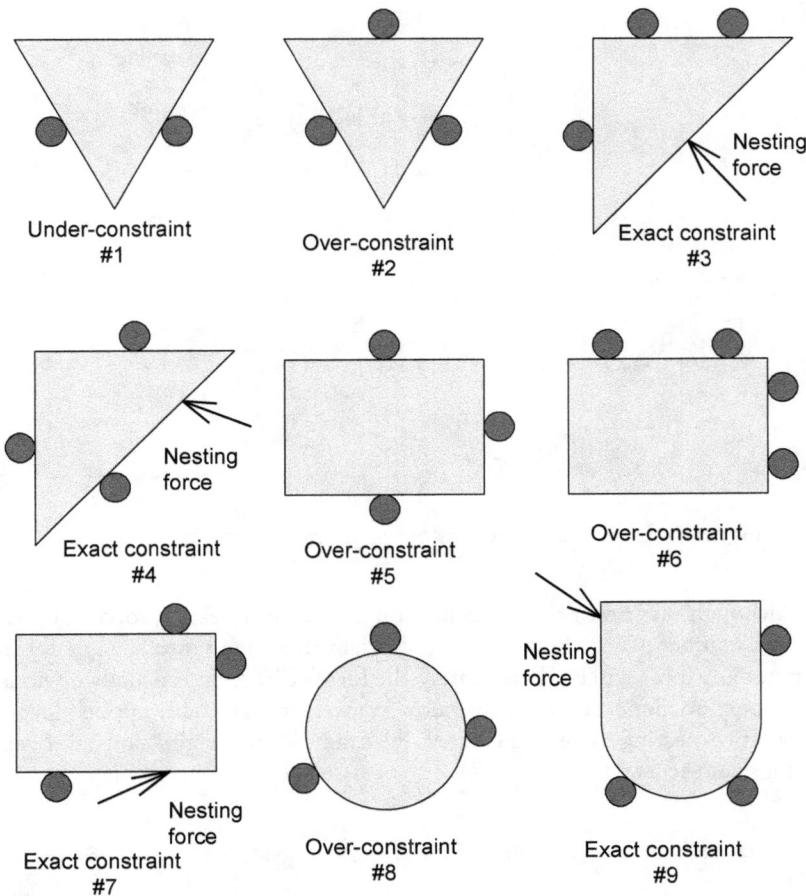

Figure 3-10 Two-dimensional constraint condition examples.

3.4   Nesting forces

You cannot just replace the nesting force with a fourth post. Dimensions are never perfect, and the post either interferes with or does not touch the plate (Figure 3-11).

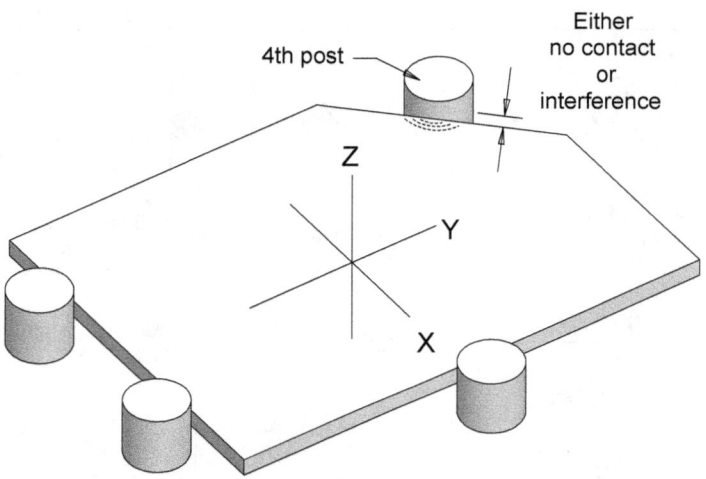

Figure 3-11 A fourth post is overconstrained and does not replace the nesting force.

Although our nesting force definition asked for a nesting force coupled to each contact point, these can be vectorially combined into a single force. But not any force will do. Fortunately, the force's direction is usually obvious by inspection alone, and its magnitude counteracts externally applied loads. In practice, nesting forces are created, for example, by weight, cams, wedges, springs, and screws (Figure 3-12).

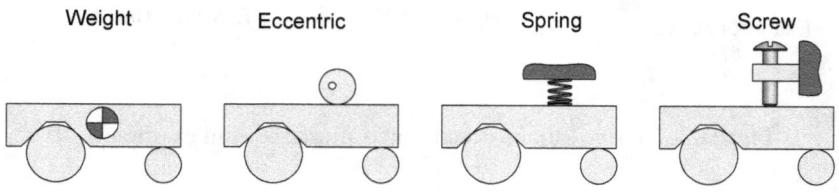

Figure 3-12 Example means of applying a nesting force. (Adapted from Blanding [12].)

There is a nesting force "window" outside of which any nesting force will tumble the component out of the stable, nested position (Figure 3-13). Appendix B shows the graphical technique to find this window.

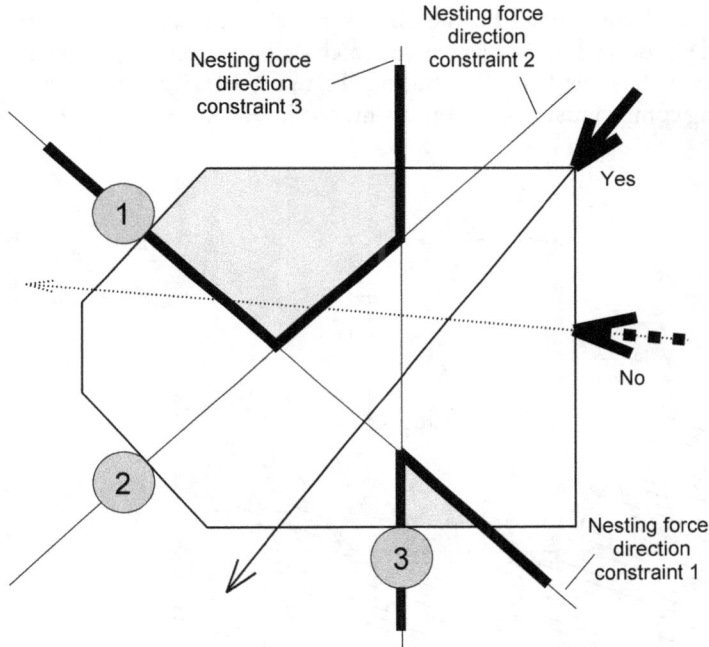

Figure 3-13 The nesting force window (unshaded area) in an exactly constrained block showing suitable (Yes) and unsuitable (No) nesting forces. Any force directed outside this window will not maintain contact at all constraints.

3.5   Constraint theory in practice

If materials were inelastic and unyielding, and, therefore, components neither deformed nor failed, we could design everything with perfect exact constraint. But they are not, and we cannot, so we compromise. Although I warned, "Never overconstrain a design," I must add, "unless you know how."

Useful compromises to exact constraint are pinned and bolted connections, ball and tapered bearings, frictional constraints (e.g. the ball and socket joint of Figure 2-2), and in-situ adjustments of over-constraint. Two important departures earn further description: curvature matching and elastic constraint design.

3.5.1. Curvature and surface matching

The trihedron receptacle of Figure 3-5, although exactly constraining three translations, creates infinite stresses at the points of contact, as does any point contact. Furthermore, the trihedron geometry is unpleasant for the shop. Instead, you can use a conical hole for circle contact to better distribute the load. Another step forward is to match or, better yet, slightly mismatch the surfaces for best load distribution (Figure 3-14). Curvature or surface matching compromises exact constraint, but improves load-carrying.

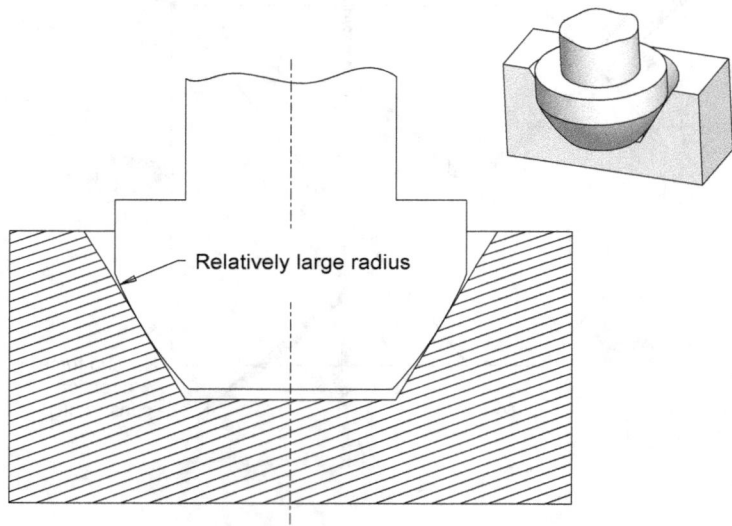

Figure 3-14 Curvature matching. This compromise to exact constraint distributes loads over larger surface areas.

3.5.2. Elastic constraint design

By now you know that four-legged chairs and tables are overconstrained. But four-legged tables wobble only if the legs are rigid, pedestal-like arrangements (Figure 3-15). With legs on the table's corners, the table top twists just enough for the fourth leg to contact (Figure 3-16). This is called, variously, elastic constraint, redundant constraint [8, 10], elastic averaging [11], and elastic design [7], and it increases load-carrying and improves stability. To use elastic constraint, combine a flexing feature with the proper nesting forces, and the redundant constraint then also contacts its mating surface.

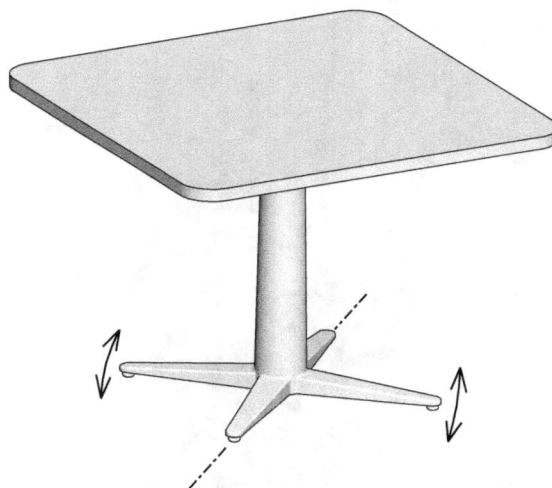

Figure 3-15 Table with rigid pedestal is overconstrained. These tables often teeter on two feet.

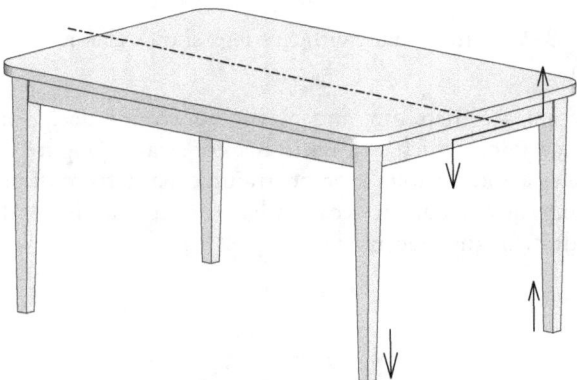

Figure 3-16 Table with four corner legs shows elastic constraint. The table top twists so all legs contact the floor, even when uneven.

If you must overconstrain a design to increase load-carrying or improve stability, you must mitigate that over-constraint with elastic deformation.

"Increase rigidity by adding flexibility?" you ask. Sounds paradoxical, but you will apply the flexibility only in the direction needed to take up variations such as manufacturing tolerances.

Office chairs never have just three casters, recent popular designs having five, which adds stability and makes them safe to lean back in (Figure 3-17). Despite the pedestal arrangement, the feet flex enough for all five to contact the floor.

Figure 3-17 Office chair with five legs shows elastic constraint design.

Combine surface matching and elastic constraint design and you get a bolted flange joint, which is a frictional constraint (Figure 3-18). Large clearance holes avoid obvious overconstraint, even with multiple bolts, and manufacturing the two flat surfaces to adequate accuracy is not difficult. The bolts easily draw in any inaccuracy.

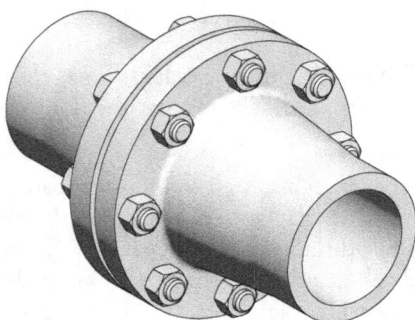

Figure 3-18 Flanged joint: an elastic, surface-matching, frictional constraint.

4. Plan the load path in parts, structures, and assemblies.

Whenever you design a part or assembly, you should clearly define the paths of applied and internal loads. First off, just visualize the load path through the part or mechanism.

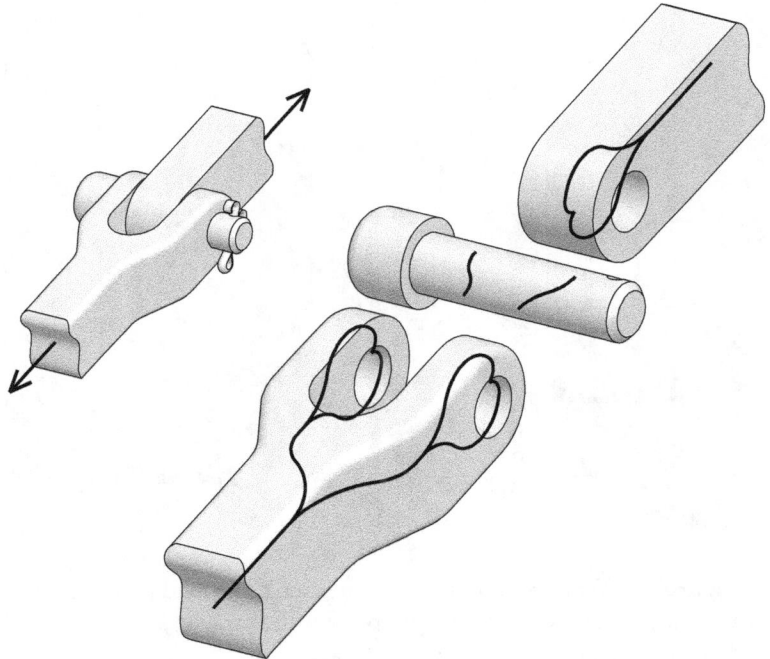

Figure 4-1 Load paths through a pinned clevis connection. The first step in planning the load path is visualizing it.

You want the load path to be:

+ Short
+ Direct
+ In a line, or, barring that, in a plane
+ Symmetric
+ Non-redundant, or, barring that, elastic
+ Locally-closed
+ Easily analyzed (if needed)

My favorite comparison of poorly- to well-designed load paths is wall-mounted liquid soap dispensers (Figure 4-2). In the pull lever version, load transfers through a lever stop, the housing, a wall plate, screws, wall anchors, and, finally, to the wall. (These dispensers are sometimes found on the floor.) The push-type dispenser transfers load more directly to the wall and over a large area, a much better load path.

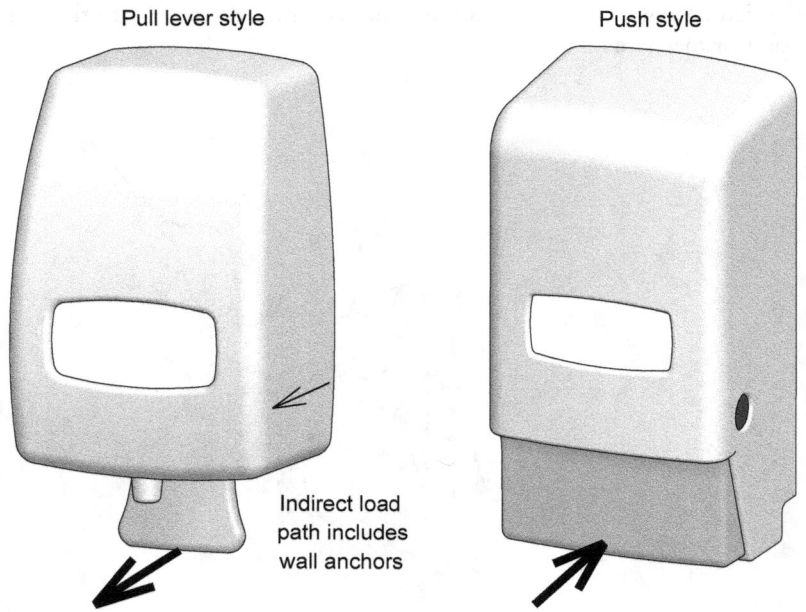

Pull lever style   Push style

Indirect load path includes wall anchors

Figure 4-2 Wall-mounted soap dispensers. The load path in the push style travels more directly to the wall surface.

Locally-closed load paths are especially preferred. Helical gears run smoother and transmit higher torque than spur gears, but develop axial thrust (Figure 4-3). Thrust bearings usually oppose this, but a herringbone configuration eliminates external thrust altogether (Figure 4-4). The thrust load paths reside entirely within the gears.

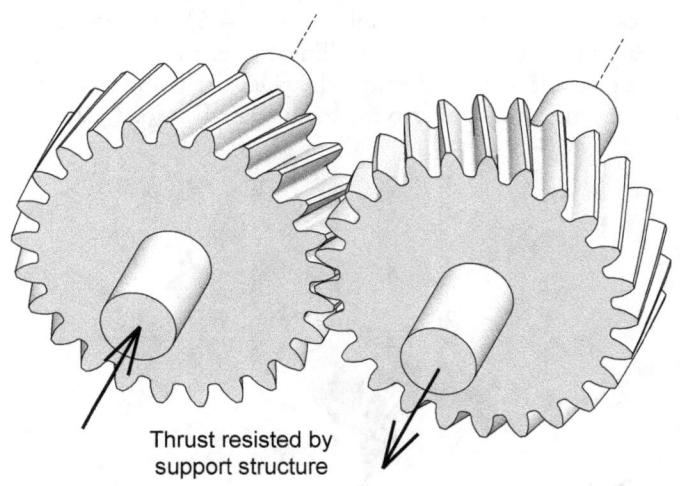

Thrust resisted by
support structure

Figure 4-3 Helical gears develop thrust, which is opposed by the assembly, usually thrust bearings.

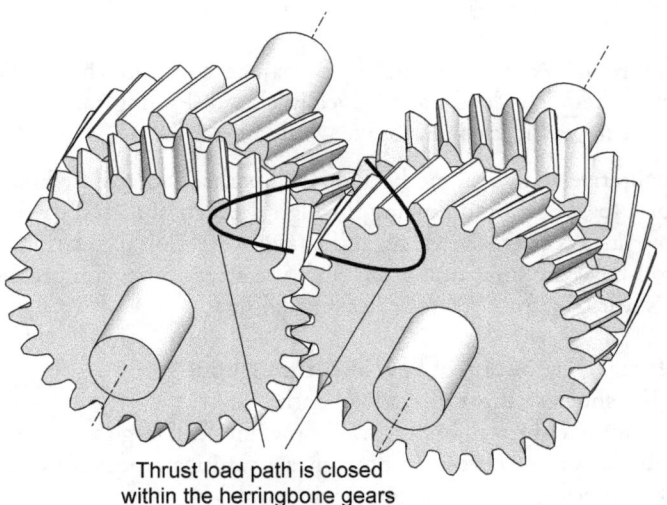

Thrust load path is closed
within the herringbone gears

Figure 4-4 Herringbone helical gears: a locally-closed load path. This design eliminates axle thrust and requires no thrust bearings.

User-centered design benefits from locally-closed load paths. Bicycle hand brakes are a common example (Figure 4-5). Always consider how user-generated forces are opposed. Pulling, rather than squeezing the brake lever would give a load path through the rider's arm and body, to the seat, through the frame, and, finally, to the handlebar. This load path is unstable.

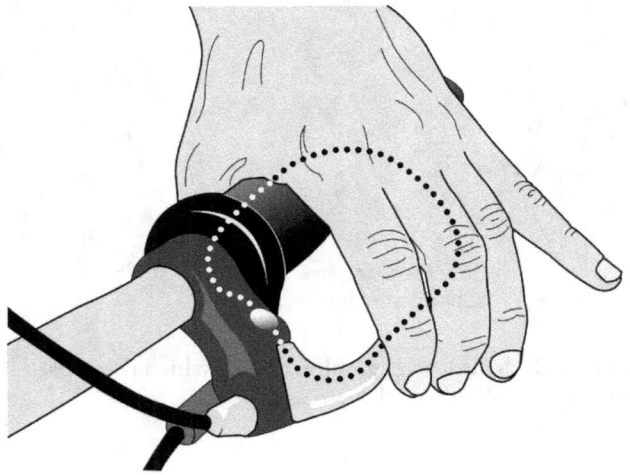

Figure 4-5 A bicycle hand brake is squeezed rather than pulled or pushed. The load path is locally-closed. Pulling rather than squeezing a bicycle hand brake would be perilous.

A qualitative analytical method for load paths is force flow line analysis. If you plot force as a fluid flowing through the structure and the joints, the fluid flow lines converge, diverge, and change direction in relation to stress. Once you have the lines drawn in, label the stress type: compression, tension, or shear (Figure 4-6). Your design goals are:

- Equal distribution of flow lines throughout
- The shortest, most direct load path
- Minimize the number of times flow lines converge and diverge
- Avoid concentrations of flow lines, or mitigate possible failures (e.g. high performance fasteners or materials)
- Gentle rather than abrupt changes in direction
- Avoid bunching around corners

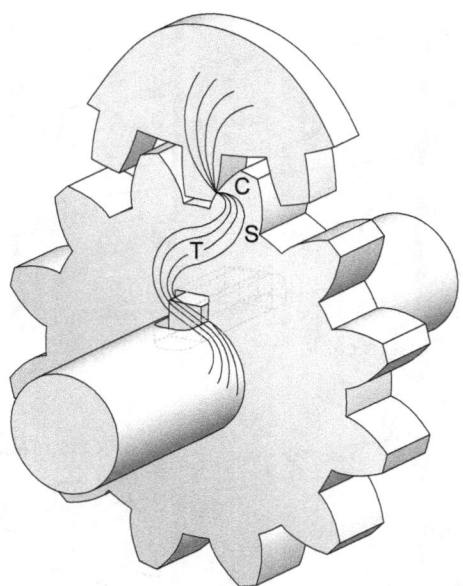

Figure 4-6 Detailed force flow lines in a gear set and keyed shaft. The lines can be labeled for tension, compression, and shear.

If you design plastic parts, pay particular attention to load paths. Thermoplastics endure neither concentrated stress nor constant loads, at least not very well. Avoid sharp corners altogether and distribute loads over large surface areas for successful plastic part design.

5. Triangulate parts and structures to make them stiffer.

Triangulation applies to structures and structural elements. When components or structures need to be stiff or strong, create triangles.

A common example of the effectiveness of triangulating for strength and stiffness is a swinging fence gate (Figure 5-1). Triangulating members prevent (or correct!) sagging. A tensile cable connects the lower outer corner to the upper inner corner, and transmits load to the gatepost. Alternately, a compression brace can be added to the diagonal to transmit the load to the gatepost.

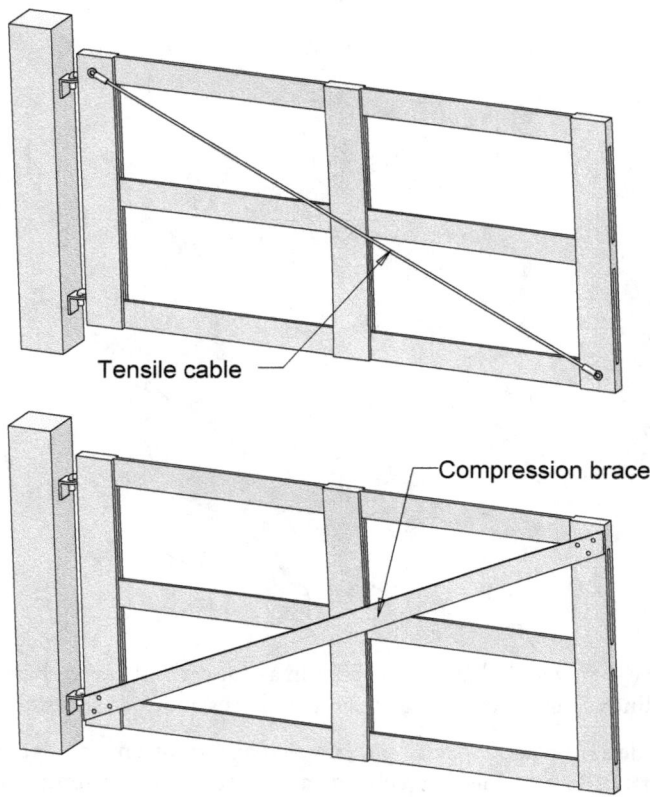

Figure 5-1 Swinging gates with triangulating tension and compression members. Structures without triangulating members rely on the rigidity of connecting joints between members for stiffness.

Also effective as triangulating members are shear webs such as a thin sheet affixed to the back of a bookcase (Figure 5-2). If you have ever constructed one like this, you will have seen that it is quite wobbly side-to-side without the back. But once the back is attached, the bookcase becomes remarkably rigid and stable in side-to-side loading.

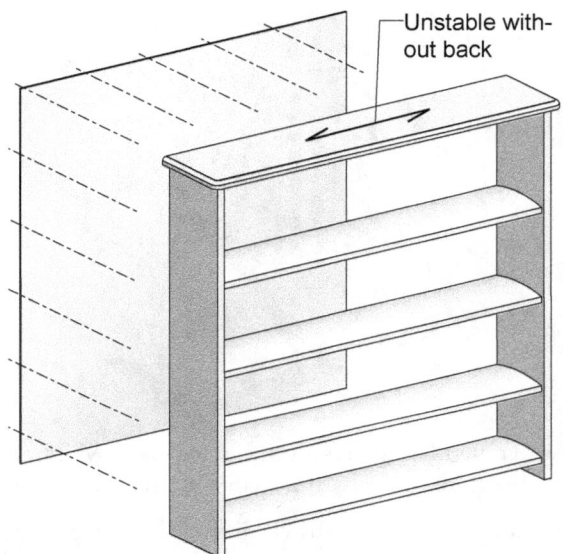

Unstable with-
out back

Figure 5-2 Bookcase with triangulating back. This bookcase will not collapse side-to-side, but still could twist corner-to-corner, an uncommon loading.

A triangulating feature need not be internal to the structure. Stiffening open boxes is a stubborn problem. In order to keep access to the box, the opening cannot be covered. One solution is to add a flange to the exterior of the rim (Figure 5-3).

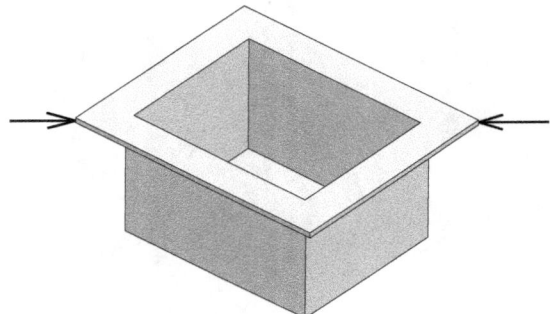

Figure 5-3 Open box with stiffening flange.

Ribs are common in molded and cast parts. They efficiently stiffen the structure by using thin braces, which are required by many manufacturing processes (Figure 5-4).

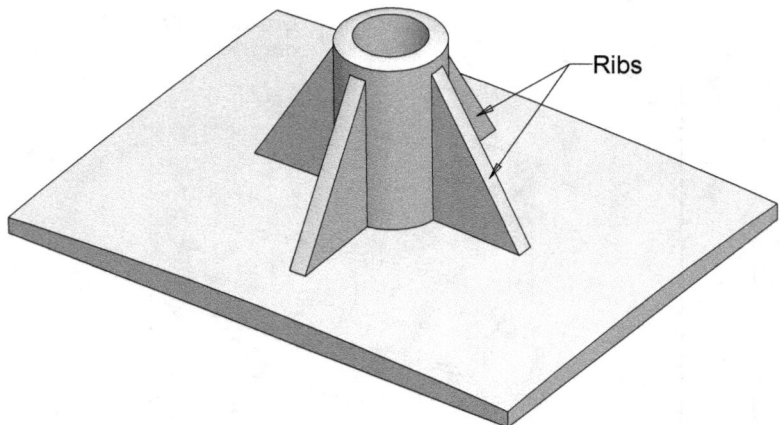

Figure 5-4 Triangulating ribs in a molded part. Molding and casting often require thin ribs and webs for strength or rigidity.

Elements that triangulate are webs (as in the bookcase example), truss members (as in the gate example), ribs (as in the boss), and flanges (as in the open box). All are tensile or compressive load-carrying members in the plane of deflection. To triangulate, first identify the plane in which the load and the undesirable stress or deflection lie, then add integrally attached ribs, webs, flanges, or truss elements in that plane.

The three-dimensional equivalent of a triangle is a tetrahedron (Figure 5-5). Create tetrahedrons, which are just four triangles, to stiffen three-dimensional structures.

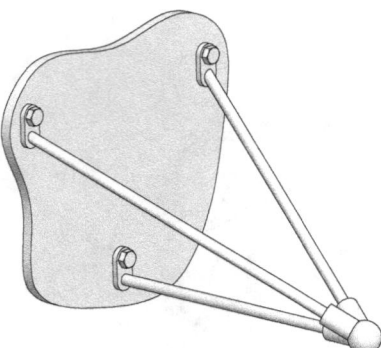

Figure 5-5 A tetrahedron: three-dimensional triangulation. Four triangles give three-dimensional rigidity.

But beware! Triangulating members and shear webs stiffen structures, but stiffer does not always mean stronger or more robust. Stiffening transfers loads to a different place, a place that might be weaker or more susceptible to fracture, fatigue, corrosion, or whatever else leads to failure.

6. Avoid bending stresses. Prefer tension and compression.

You might recognize that this statement repeats, more-or-less, the previous section. Triangulating members typically have tension and compression loads rather than bending loads. But stating this anew offers a different insight into structures.

Bending produces stress distributions that vary from zero to some maximum (Figure 6-1). Similarly, torsional stress varies from a maximum in the outer fibers to zero in the center of the section. On the other hand, pure tensile or compressive members have constant stress throughout (Figure 6-2).

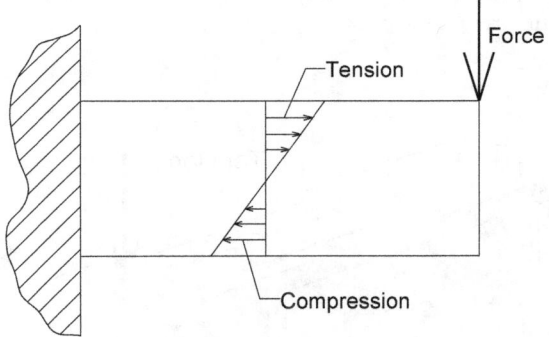

Figure 6-1 Stress distribution in a beam under bending load. Material at the midpoint of the section is unstressed and contributes nothing to strength or rigidity.

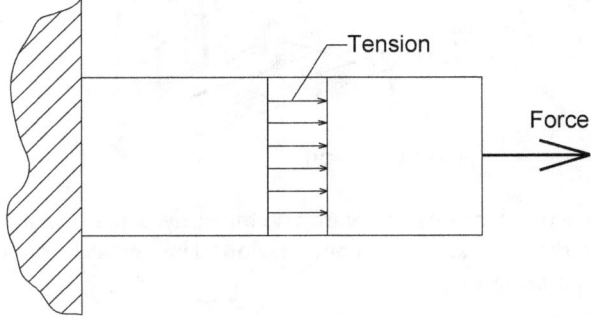

Figure 6-2 Stress distribution in a beam under tension. All material contributes to carry load.

In bending and torsion, much of the material contributes nothing to carrying load. The highly stressed areas fail even while the neutral axis material remains unloaded. Clearly, bending and torsion loads are inefficiently born by structural elements, whereas uniformly-loaded elements use material quite efficiently.

So another way of saying this is:

Design to get uniform stress [26].

Even when this ideal is impossible, you can approximate it. I-beams are a practical way to use material efficiently to support bending deflections. Most of the material is at the maximum possible distance from the centroid. Thus, the material in an I-beam is nearly uniformly loaded, and at the maximum stress (Figure 6-3). But there is more to an I-beam. When viewed in three dimensions, the internal connecting section is a triangulating shear web as described in the previous section.

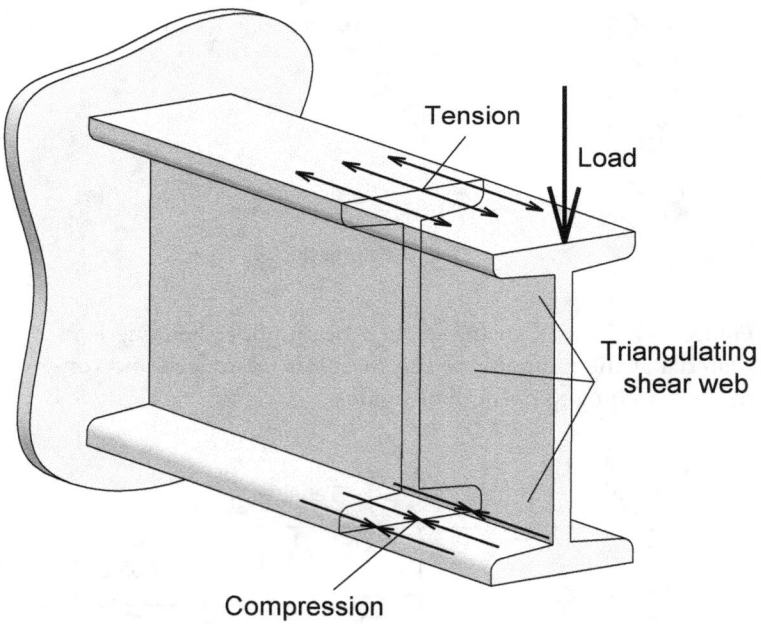

Figure 6-3 I-beams use material efficiently. The horizontal portions resist tension and compression. The vertical portion is a triangulating shear web.

For torsional loads, an analogous statement is to avoid transverse or torsional shear, and to prefer uniform shear. Therefore, the lightest weight torsion shafts are thin wall tubes (Figure 6-4).

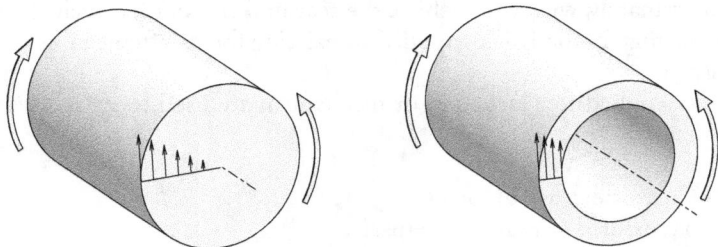

Figure 6-4 Shear stress distributions in solid and hollow torsion shafts. Material near the center of a solid shaft carries no load. Material in a thin, hollow shaft has nearly constant shear.

Of course, you will sometimes want to take advantage of bending and non-uniform torsion. Helical compression springs are torsional beams with non-uniform shear. Likewise, torsional springs are cantilever beams loaded in bending. Consider snap-fit features; these are cantilever beams loaded in bending (Figure 6-5). These examples suggest the advantage of non-uniform stress: flexibility. Any time you seek flexibility in a part or structure, bending or non-uniform torsional loads are the better answer.

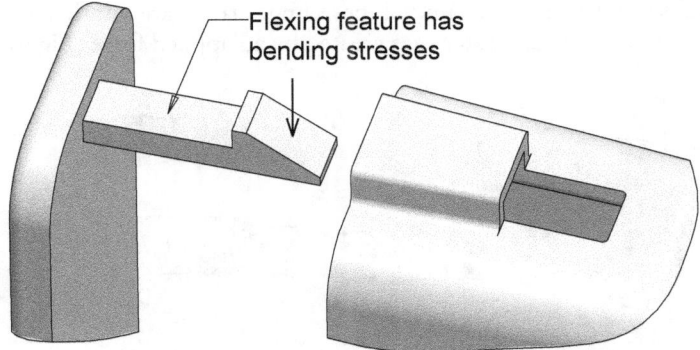

Flexing feature has bending stresses

Figure 6-5 Bending is an advantage for parts requiring flexing: a cantilever snap fit.

7. Improve designs with self-help.

Why do your vehicle's front brakes do more braking than the rears? Because braking itself transfers weight to the front wheels improving their grip on the road. This is self-help, which simply stated is:

Use applied loads to improve performance.

Unfortunately, with your vehicle the rear brakes work less well. They are self-damaging. Nonetheless, applied or existing forces often can be to your benefit.

Forces applied to a structure or mechanism are used to great advantage when they:

+ Create new, useful forces
+ Transform or redirect themselves
+ Balance either themselves or existing loads
+ Help to distribute loads

7.1   Self-help that creates forces

Scouts and mariners know that square knots hold and granny knots don't, and that bowlines are better knots still. Good knots have force-creating self-help wherein applied loads produce stable locking friction: the higher the rope tension, the higher the locking action. Pressure-formed seals such as O-rings and internally-mounted doors (e.g. fermentation tanks, autoclaves, some airliners) use self-help to create robust seals. The capstan and windlass use self-help to pull in boat anchors by creating and multiplying friction through several turns of chain. Chinese finger traps are another common example in which useful forces are created by an applied force (Figure 7-1).

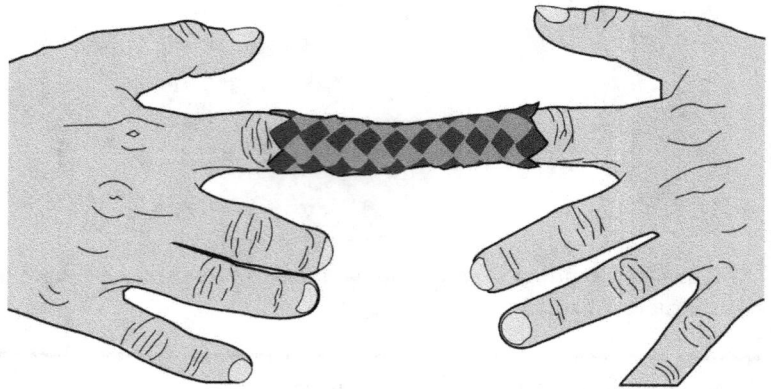

Figure 7-1 Chinese finger trap—an example of self-help that creates forces. The harder you pull, the tighter it grips.

7.2   Self-help that redirects forces

My favorite example of self-help that redirects an applied force was offered to me by a left-handed colleague. He explained that left-handed scissors have little to do with how thumb and fingers fit, which I erroneously thought, and lots to do with how hand action naturally forces the blades' cutting edges together (Figure 7-2). This is a vivid lesson in designing forces to help rather than hurt performance.

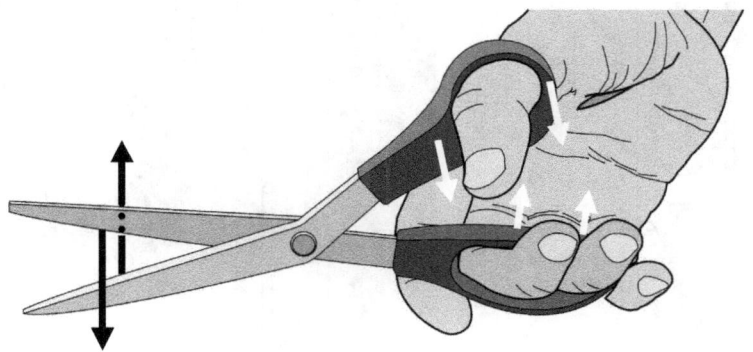

Figure 7-2 Self-help in a scissors. Normal hand action forces the blades' cutting edges together. Left- and right-handed scissors are mirror images of each other, improving the scissors' shearing action for either user.

7.3   Self-help that balances forces

Self-help can balance or neutralize, at least partially, an undesirable effect. Common examples are counterweights on draw and lift bridges and in cable-operated elevators. Although counter-weighting increases working weight, it eliminates moments in the structure and reduces actuator power.

Some stadiums and hotels have special doors, called balanced doors, that use balancing self-help to keep them closed in wind, yet easy to open or close, even in strong wind. The hinge point, rather than being at an outer edge, is inboard when the door is closed. Wind pressure on the door face is balanced on either side of the hinge axis (Figure 7-3). As the door opens, the pivot axis slides by mechanical action to the side to open the door fully.

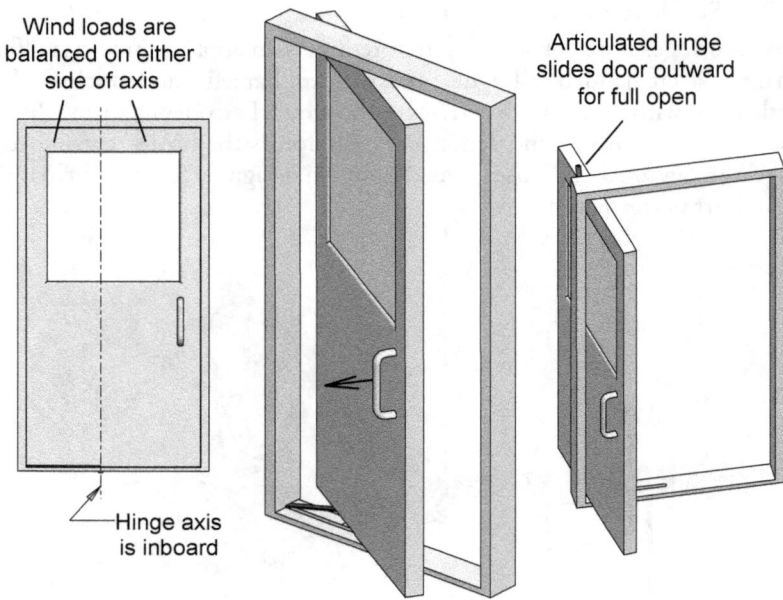

Wind loads are balanced on either side of axis

Articulated hinge slides door outward for full open

Hinge axis is inboard

Figure 7-3 A balanced door with an articulated hinge exhibits self-help. Wind will not open this door, yet it opens easily in strong winds.

These examples balance like forces: weight with weight, wind load with wind load. But you can also balance unlike forces. Examples often cited include tanks in which pressure and thermal loads offset, rather than add to, each other and gas turbine blades with an intentional tilt so that blade stress from centrifugal force partially offsets, rather than adds to, bending stress from gas impingement.

7.4   Self-help that distributes loads

In this case, the load path changes with deformation, thereby improving performance. A graphic example is Hertzian contact. Ever more material is recruited to withstand the load as that load is applied (Figure 7-4).

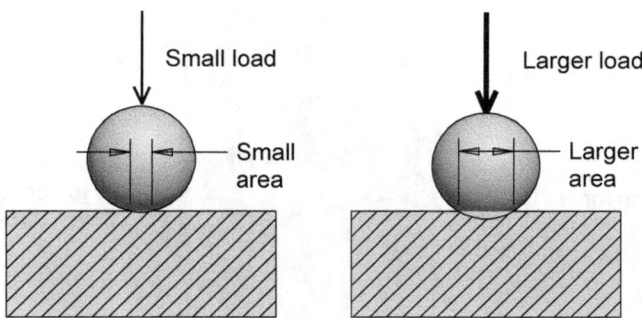

Figure 7-4 Hertzian stress: simple load-distributing self-help. The larger the load, the larger the contact area.

Physical stops to prevent overtravel are a rudimentary example of load-distributing self-help. But structures can also be designed to stiffen or strengthen with deflection. For example, beams become much stiffer with large deflections (if they don't yield) as the bending progresses to tension.

8.　Manage friction in mechanisms.

The only thing certain about friction in mechanical design is that you will have it. How much, and its consequences, is uncertain. Never trust friction—manage it.

8.1　Avoid sliding friction.

Experienced designers avoid sliding friction whenever possible. Haven't we all struggled with enough sticking desk drawers to know that? But there are scientific reasons why managing sliding friction is difficult: stick-slip behavior owing to the difference between static and dynamic friction, wide variation in frictional coefficients, uncontrolled lubrication status, variable surface finishes, surface damage such as galling, and wear.

Sure, sliding elements are tempting, and there are times, driven by cost, that you will compromise. But if you do, select suitable materials, carefully design the guiding system, test beyond reason, and expect the worst.

8.2　Maximize the length of linearly-guided components.

The best way to avoid a sticky drawer is to design it long and narrow. Linearly-guided components jam when the insertion force is inadequate to overcome the friction of the guides' contacts (Figure 8-1). Short, wide systems are especially susceptible not only because of the length-to-width ratio, but because the insertion force is too easily applied off center.

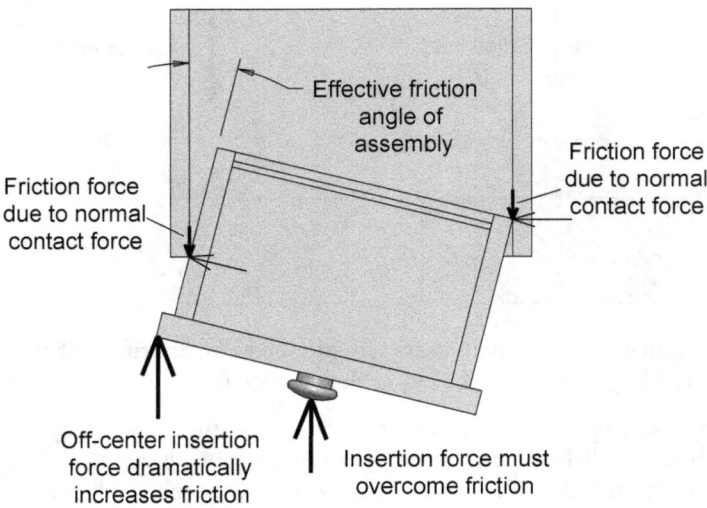

Figure 8-1 Diagram of friction angle and forces in a short, wide "sticky drawer."

The friction angle at lock-up depends on the coefficient of friction, the drawer's length and width, and the insertion force's offset from the centerline.

You want as long a "wheelbase" as space will afford, and you want applied loads as close to the centerline as possible (length-to-load arm). A good minimum rule-of-thumb for length-to-width and for length-to-load arm is 1.5 to 1, but don't trust that without testing it. Furthermore, your design must guarantee contact only at the ends and not in the middle of the moving component, which would only reduce the effective wheelbase. Relieving the middle section in solid parts is one solution, as is using two end-mounted bearings.

Here is an important additional note about linearly-guided systems and exact constraint: linear systems rigidly guided on two separate axes are over-constrained (Figure 8-2). These will jam and stick and wear out—and be difficult to manufacture. Solutions are not so obvious and are often painful, but eliminate the fixed center-to-center distance in either the carriage or the rails, or expect trouble.

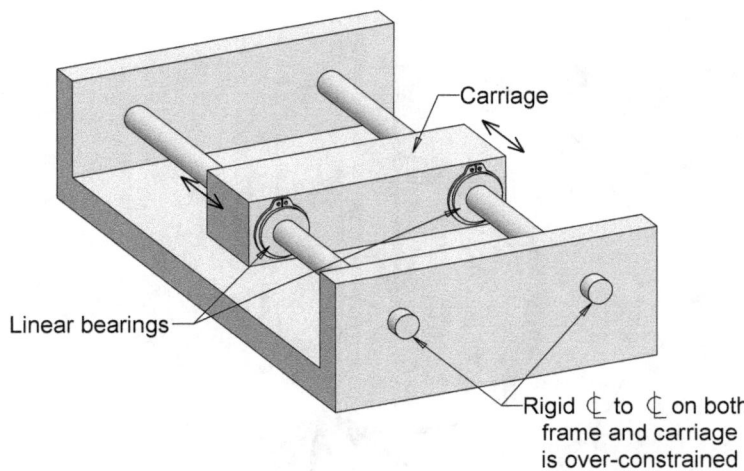

Figure 8-2 A poor design for linearly-guided systems. The center-to-center distance is fully constrained two different ways, giving over-constraint.

8.3   Select rotary motion over linear motion.

I have often read, and once believed, that the wheel is the most important human invention. But while researching for this book I came to believe that the most important invention is not the wheel, but the journal bearing. It is why rotary is so much better than linear motion.

I am unable to imagine any but clumsy uses for wheels alone. OK, moving large stone blocks over the ground using log "wheels" probably had its day, but applications for this are few. But put the wheel on an axle, the axle in a journal bearing, and you leap ahead enormously. Why? Because:

$$M = R \times r \times \sin \phi = F \times r$$ (Equation 8-1, reproduced from Beer and Johnston [32])

where M is the moment to turn the axle, r the radius of the axle, W the load, and R, F, and N the reaction forces (Figure 8-3).

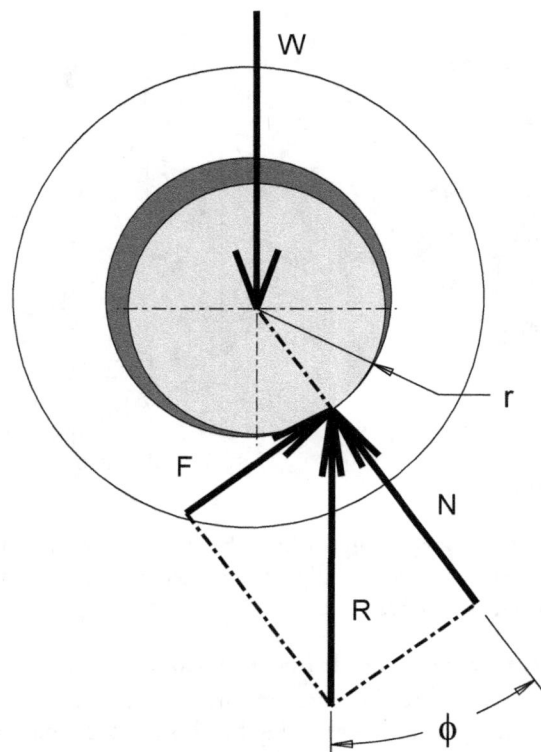

Figure 8-3 Forces on a simple journal. (Figure 8-3 and Equation 8-1 are reproduced with permission of the publisher from Beer and Johnston, *Vector Mechanics for Engineers, Statics and Dynamics, 5th Edition,* ©1988, The McGraw-Hill Companies [32].)

In this relationship, we should treasure the axle half-diameter, r. True, ϕ is governed by the coefficient of friction (the lower, the better), and lubricants have a useful place in this as well. But the real power lies in r. Make it zero, and you have zero friction!

To "Select rotary motion over linear motion," add:

To reduce friction, use small diameter axles.

History offers a keen example. John Harrison, an 18th century English clockmaker, and creator of the first navigational chronometers, is also credited with inventing the caged roller bearing. Reducing friction for his chronometers is what led him to this invention (Figure 8-4) [34]. Roller and ball bearings have advantageous rolling friction rather than sliding friction—unless the rollers or balls touch each other. A cage separates the rolling elements in Harrison's design [35], and in modern bearings as well. Rotational sliding friction between cage and roller is still unavoidable, but the miniature axles on the rollers manage friction by using the power of a small r in Equation 8-1.

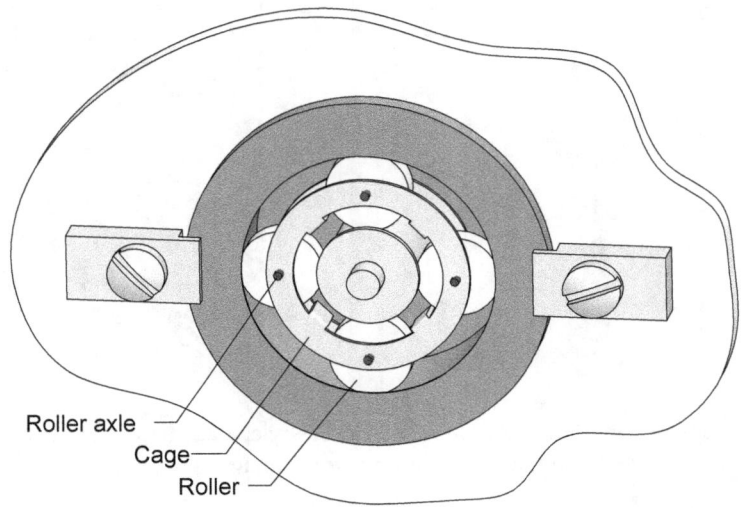

Roller axle
Cage
Roller

Figure 8-4 John Harrison's caged roller bearing. (Adapted from Andrewes [35].)

But the history lesson continues. Harrison used jewel rather than roller bearings in his later chronometers because of their smaller size, even though rolling element bearings are thought of as frictionless, and jewel bearings have rotary sliding friction. But jewel bearings were quite sufficient at reducing friction, fundamentally, again, because of the relationship in Equation 8-1.

Of course, it is only a small extension to apply the concept of Equation 8-1 to thrust bearings. To minimize friction, thrust bearing surfaces should be at the smallest possible diameter as well.

In large part, it is the relationship of Equation 8-1 that makes rotary motion much the better selection over linear motion in mechanical design. Of course, linear motion works with proper guiding, and with good linear bearings, but the helpful "amplification factor" of this relationship never exists in linear motion.

Also in large part, it is the friction angle relationship of Figure 8-1 that makes linear motion much the worse choice over rotary motion. In order to overcome this relationship, you will need space, and lots of it, for linear guidance. Rotating bearings, of whatever form, offer stable, compact guidance to the moving components of a mechanism in a minimum of space. Figure 8-5 shows an everyday example.

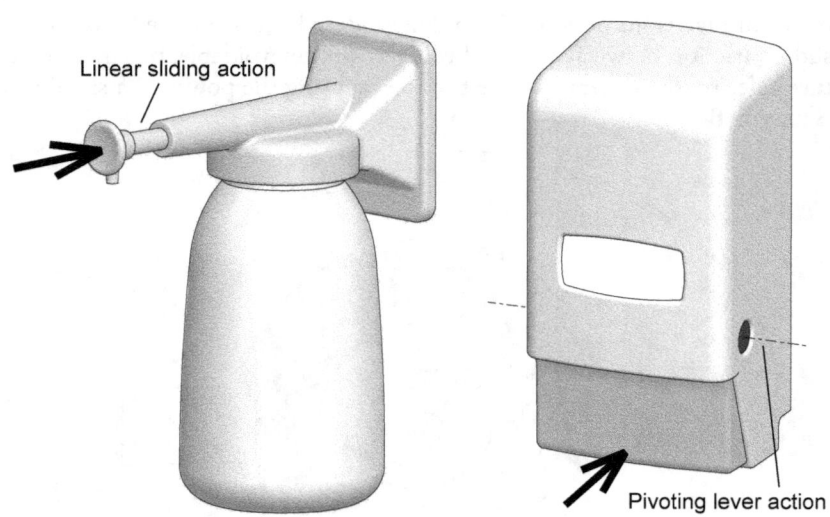

Linear sliding action

Pivoting lever action

Figure 8-5 Plunger style versus lever-style soap dispenser. The lever-style has a natural advantage for managing friction.

8.4   Use rolling element bearings whenever possible.

A reliable way to manage friction is to use rolling element bearings. Furthermore, rolling elements often eliminate wear. Ball bearings, roller bearings, needle bearings, tapered roller bearings, and recirculating ball linear bearings are wonders of mechanical engineering and modern technology.

A great advantage of rolling element bearings is that they help to keep it simple; the complexity of ball bearings—and they are complex—is intrinsic, and doesn't normally concern designers. You only have to use specifications and application notes, which are remarkably reliable.

I have more than once relearned that rolling element bearings are not expensive, despite a widespread prejudice that they are. Alternatives such as sleeve bearings, which have a place, often require difficult assembly, tight tolerances, secondary machining, and meticulous design, and, most important, do not reduce friction and wear nearly as reliably as rolling element bearings.

Unfortunately, however, with bearings you are walking a tightrope above over-constraint. Simple ball bearings have the inherent advantage, over sleeve bearings, of small rotational degrees of freedom to accommodate some misalignment. Beyond that, keep a watchful eye out for over-constraint and mitigate with self-aligning and adjustable features.

8.5   Use flexures to eliminate friction.

For a completely frictionless system, you cannot do better than flexures. These are uncommon, specialty components, and are typically reserved for precision instruments and the like. Nevertheless, consider using flexures for precise, frictionless motion over a limited deflection (Figure 8-6). The lack of friction means that flexure-defined motion is predictable and reliable, unlike friction, which is neither.

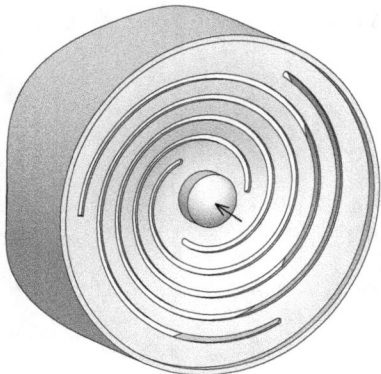

Figure 8-6 Triple spiral frictionless flexure for guiding axial displacement.

PART II

ESSENTIALS OF THOUGHT AND PROCEDURE IN MECHANICAL DESIGN

9. Use three-dimensional solid model layouts to find the best arrangement of parts and assemblies.

Engineers and designers use graphical representations for two important purposes:

1. As an aid to thinking
2. For communicating design ideas

There are two traditional types of graphical representation: orthographic projection and perspective drawing. Three-dimensional (3D) solid models have become a third type of graphical representation, and have largely replaced the other two for designing mechanical assemblies, both as a thinking and a communication tool.

Let's be truthful. 3D solid modeling is how mechanical design is done. Not every step of the way, and not by everybody, but it is the predominant tool for defining geometry during mechanical design. The advantages are overwhelming: exceptional visualization, continuity of effort, and simple, unambiguous communication.

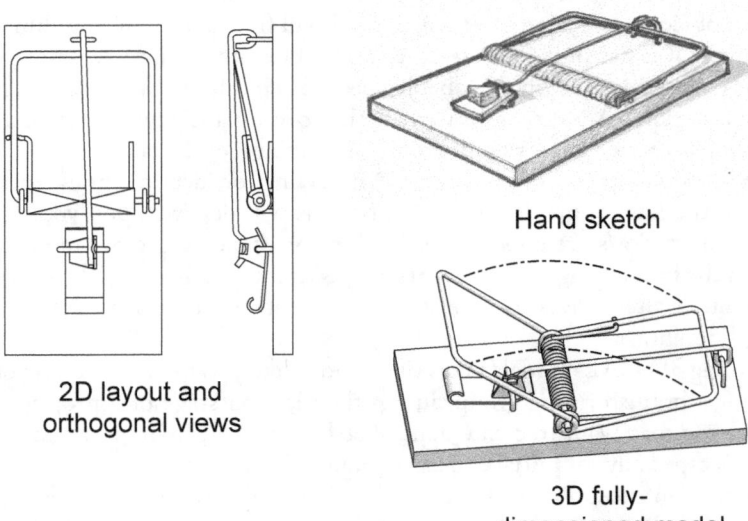

Hand sketch

2D layout and
orthogonal views

3D fully-
dimensioned model

Figure 9-1 2D layout, 3D sketch, and 3D model. A fully-dimensioned 3D model captures more information than the other two.

So why has 3D modeling been embraced by the mechanical design community? For the same reasons that graphical representations have always been useful, but even more so. Everyone's visualization skills and memory, no matter how good, have a limit. Graphical representations supplement your short term memory and allow you to build upon what you already have [36]. 3D solid models do this dramatically better, and you cannot afford to forgo this advantage.

To design a mechanical assembly, make a 3D layout. Start with simple representations of the parts, assemble them together, and check for fit and function. If you haven't learned 3D modeling, work with someone who has. This layout is an essential representation of a mechanical assembly during design. Why?

Scale. It is the rare designer who can grasp, without the aid of scale representation, all of the constraints, limitations, and possibilities of a design concept. Often, a promising idea simply does not work because there is no room for all the parts and functions. Furthermore, few design projects start with completely clean sheets of paper. More often, we must design around—and fit into—what already exists. Most of us need help, and nowadays, 3D solid models provide it.

But there are caveats and exceptions!

One obvious exception is simple parts. I will freehand draw and dimension a part if it has only a few features and it does not go into an assembly. (If it is part of an assembly, I probably need a 3D model anyway to check for fit and function.) Another is for parts with predominantly two-dimensional characteristics such as extrusion profiles, stampings, or die cuts.

An accepted caveat is not to use 3D modeling for ideation, brainstorming, and the like. Some do, some do not; this will depend upon your and your organization's preference. This book recommends using 3D modeling "to find the best arrangement of parts and assemblies." How initial concepts and multiple alternatives are identified, documented, and presented is something different.

Making 3D CAD models takes time, and is likely to be slower than your brain or a brainstorming group during the ideation stage of a project. So indeed, you may want to create hand sketches, both orthographic and perspective, especially for initial concepts (Figure 9-2).

There is an exception to this exception. You may have to use 3D layouts as part of the concept selection process. How can you decide among different designs without understanding their geometric constraints? I have more than once selected the "best" design from among different options, only to find intractable space or size difficulties with a 3D layout, so this can be a vital part of the screening process.

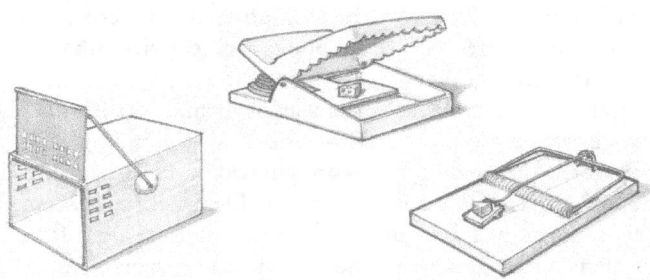

Figure 9-2 Sketches are valuable when generating multiple concepts. 3D CAD models are often unnecessary at the early ideation stages of a project.

You may prefer to sketch at least a little before starting a 3D model. A sketch can suggest a starting configuration, one that will prevent wasting 3D modeling time. But most designers of mechanical assemblies jump to 3D solid models quickly, sometimes right from their mind's eye. Investing too much time in a hand sketch wastes time. You will need a 3D model before long anyway, so once you are committed to a direction, invest in the model instead.

But you must have discipline when creating 3D models. Do not create any more detail than is necessary at each stage—this wastes more time than you can afford. Functional prototypes do not need the same level of detail as a final molded part. Test it first, add detail later.

Finally, never let a design's history decide its future. Having invested a small career in a 3D model is not sufficient reason to use it. "It is too much work to redo," will not justify a substandard design. If there is trouble, fix it. If you have to scrap it, scrap it. Chalk it up to experience, and don't make the same mistake next time.

10. Invert geometry to reveal new solutions.

Mechanical design is all about geometry, a matter of organizing features and components in three-dimensional space. Shouldn't be too hard, right? Most devices have a few requirements or a few basic functions. Work these out and you should be done. But then, unfortunately, there are design constraints to deal with. Above all in detail design, these are limitations to size and space, but material choices, environmental conditions, cost, manufacturing, and, of course, user-centered design all add to the challenge.

So how best to reconcile requirements and functions with size and space constraints? There are formal techniques for generating needs, requirements, and concepts, such as Quality Function Deployment, brainstorming, and TRIZ. Formal techniques for evaluating and selecting design concepts are also commonly used. There are books on creativity enhancement, as well as for user-centered design.

But, frankly, there is little in the way of formal methods at the more detailed levels of mechanical design—where, for the most part, a mechanical designer works out the physical configuration and details of a design. Nor is there any consensus on how to do it. Designers just seem to immerse themselves in the problem and think it through, trying different combinations, widening and narrowing their perspective, resolving or compromising issues, until the design converges.

But there is one particularly useful thought process, especially for seemingly intractable problems: inversion. Just ask yourself, "What happens if I flip things around, or over, or inside-out?" (I use the term inversion rather loosely, meaning more than just swapping one position for another.)

I try inverting a design whenever I hit a roadblock, and my predicament somehow always seems to yield to this. This works at the layout stage of design because this stage is all about manipulating geometry, and that is exactly what inverting does.

To get the idea, here is a list possibilities, and do not forget to consider the reverse of these.

inside → outside
right → left
above → below
symmetric → asymmetric
in-line → offset
smaller → larger
parallel → normal
oblique → normal or parallel
concentric → eccentric
moving → stationary
pressure → vacuum
axial → radial
flat → curved
bolt → nut
peg → hole
stiff → elastic
translating → rotating
2-dimensional → 3-dimensional
mirror about a plane

A clever example of design inversion from the medical industry is cuffed endotracheal tubes inverted from pressure to vacuum. These tubes have an inflatable annular balloon, or cuff. Once inserted, the cuff is inflated to seal the tube to a patient's trachea during anesthesia or artificial ventilation. But an inverted design, also commercially available, uses a normally-expanded foam ring covered with a balloon. Vacuum collapses the foam cuff for insertion, then the foam self-expands with vacuum release for sealing. The obvious advantages are that the balloon cannot be over-inflated and that leaks or punctures do not prevent the cuff from doing its job.

I used inversion for creating the clutch example found later in this book (section 18). My first thought, shown on the left of Figure 10-1, proved rather awkward. Designing a clean housing assembly wasn't working for some obvious reasons. But inverting the ball movement from axial to outwardly radial made the housing design much better (Figure 10-1, on the right). It should. That is how most commercial clutches of this sort are designed anyway!

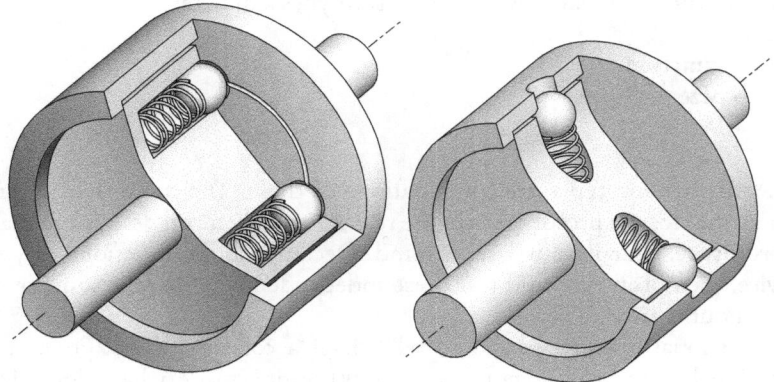

Figure 10-1 Spring clutch inverted from axial to radial orientation of springs and ball motion.

---

11. Build prototypes of everything—but not all at once.

A common reaction upon first seeing or handling a newly designed model, even by the person who designed it, is:

> "It's much [larger, smaller, wider, deeper, shorter, flimsier, heavier, clumsier] than I imagined!" (*Circle all that apply.*)

There is no substitute for a real, physical model or prototype. Equally important to experiencing a device's size and shape is gaining insight into its behavior. You will gain precious insight with physical models and laboratory testing.

It is tempting to complete a design, especially when using 3D solid modeling, without making physical models of any of it. You can see it all right there on the screen, and everything looks good, right? Sometimes you will get away with this, and sometimes it may even be the most efficient path. Simple fixtures, for example, can be made and debugged on the fly. But for production items, you risk delaying the project and embarrassing yourself because some fundamental feature or function isn't quite right, and you should have tried it before wasting time on the rest of the design—and on redesign.

Every book on design has definitions, descriptions, and prescriptions for the prototyping phase of development. Every person, project, and organization has different definitions for prototypes, and every prototype has a different purpose. Put simply, however, prototypes simulate:

+ Function
+ Size (or form)
+ Process

Actually, prototypes are combinations of these. The more that is combined, the closer a prototype is to the real thing. When you design and make a prototype, bear its purpose in mind. Decompose the functions of your device, then prototype and test these independently if there is some question about their viability.

Functional prototypes should tell you if a concept works. For purely functional prototypes, do not worry about how it is made, what it looks like, and how it is assembled. Although you should not worry about these things, you must consider them. Take the design ahead a little, at least in your mind, and have a possible path. You do not want to test a design that is impossible to complete or produce.

Sizing or form models should show you how a design looks, feels, and handles. For a pure sizing prototype, worry about these, and not its materials of construction, its internal function, or how it is made. But again, you will want to consider them.

Process prototypes help you evaluate a manufacturing process, and are just as important as those for function and size.

You are likely already familiar with ways of prototyping: machining, rapid prototyping, prototype tooling. But do not overlook less-sophisticated attempts at prototyping. Paper, tape, cardboard, foam core, pins, and glue go a long way to simulate structures, mechanisms, and housings. I am proud, not embarrassed, to get out my scissors and cut up paper like a kindergartener if it gives me quick insight into a structure or mechanism.

Figure 11-1 Structural prototypes from foam core board to compare 3D bracing of a torsion profile. These models can be tested for comparative performance.

By all of this I do not mean for you to overlook virtual prototypes and analytical models. These are extraordinarily valuable, and are getting better all the time. A colleague reminds me that he always makes virtual models before physical models. "Why would I want to make a physical model without first proving the physics of the design with an analytical model?" But one or the other alone is rarely enough, and no matter how good, virtual models cannot give anyone a sense of size and feel like real models do.

Much value in virtual prototypes and analytical models lies in investigating variations, not in determining absolute performance. A physical prototype ties the virtual world to the real one, and makes all of your virtual prototype variations more realistic as well.

You may protest that a prototype of your current project is impossible, and that will be true (e.g. expensive one-ups such as test or assembly equipment). But you can nonetheless assure success. Base the design on similar, successful designs. (These act as prototypes.) Make prototypes of subsystems you are unsure of. Do not use unproven technology. Finally, apply the contingency rule during design (see section 15).

12. Separate strength from stiffness—and stiffness from strength.

Strength is confused with stiffness and vice versa—even by mechanical engineers. Definitions vary, unfortunately, but no matter. Understand the difference conceptually, then design accordingly.

There are two elements to strength: material and geometry, which are unrelated. Likewise, stiffness, or rigidity, consists of the same two unrelated elements, material and geometry.

Strength is how much load causes yielding or breaking. If something fails, it is not strong enough. In mechanical parts or structures, the input is load, and failure is yielding or breaking. Materials fail when applied stress exceeds a material's allowable stress, as described in classical mechanics of materials and with stress-strain diagrams. Note that nothing in this definition suggests how much the part or structure deforms before yielding or breaking.

Stiffness, or rigidity, defines how much something deforms when load is applied. A material's stiffness is defined by elastic moduli (Young's modulus or, for torsion, shear modulus). For parts or structures both geometry and modulus govern stiffness. Note that nothing in this definition suggests failure—yielding or breaking.

You cannot compare Figure 12-1 with the familiar stress-strain curve for steel. The curve's non-linear shape exists because of the structure, not the material modulus. A non-linear stress-strain curve for a structure confirms neither material yielding nor non-linear material behavior.

It is quite possible for steel-supported floors to feel like trampolines, and wooden-joist floors to feel like bedrock. Despite a trampoline feel, it is also possible for this steel construction to support more weight, making it stronger, if less rigid. The difference is how much material is used and the structure's design.

Strength and stiffness are theoretically unrelated characteristics. Practically, nonetheless, they often track together. Most materials or structures that are stiff are strong and vice versa, but do not trust this. For example, soft, fully-annealed austenitic steel wire is not nearly as strong as full-hard Martensitic wire, but they are equally rigid. Note that both characteristics, stiffness and strength, are most useful as comparisons, and are best thought of as relative to something else.

No doubt you learned all of this already, if not exactly this way, and materials and structures are far more complicated subjects than what is written here. Fracture mechanics, creep, and fatigue all contribute to material failure, and materials, annoyingly, do not obey Hooke's law. A review is not the intention of this book, just some insight. If you need a stiffer material, you need a higher modulus, if you need a stronger material, you need a higher yield or ultimate strength.

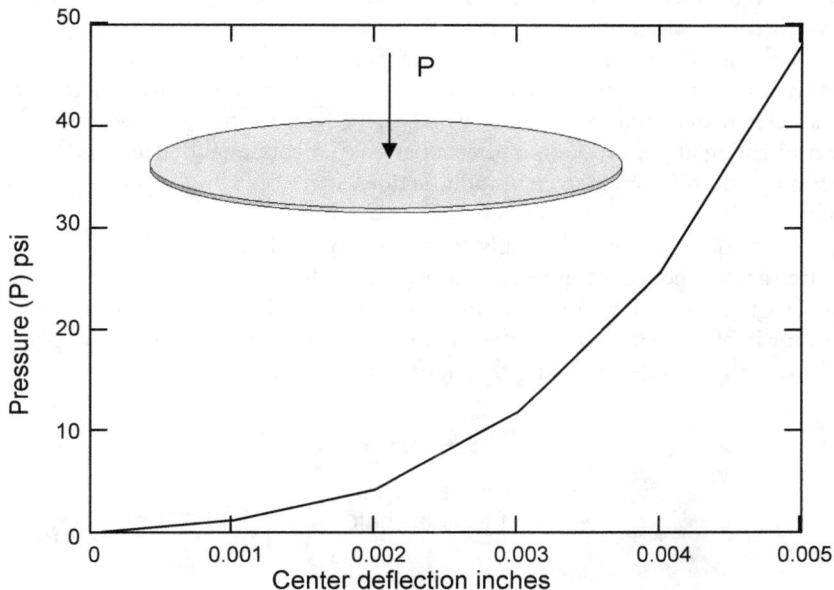

Figure 12-1 Load-deflection curve for large deflections of a flat, circular plate.

Furthermore, the most effective way to either strengthen or stiffen parts or structures is with geometry, not material. Why? Because geometry often has power relationships to exploit, whereas material improvements are often incremental. (Note: This applies to groups of similar materials. For example, a plastic part that is too weak is almost always easier to strengthen by improving the structure than by selecting a stronger plastic. Replacing plastic with steel is a different story.)

Here are two final tips:

+ Even if something won't break, do not assume it is stiff enough for its purpose.
+ Stiffening something may precipitate failures where none existed before.

13. Never overlook buckling phenomena in parts and structures.

Catastrophic failures of man-made objects are often precipitated by buckling (e.g. Hartford Civic Center Arena [43], Quebec Bridge [44]). Buckling is less well understood than other deformations, is easily overlooked, and gives little warning.

Buckling is not a material failure like yielding or breaking. (At least initially, although if unchecked, it usually results in material failure as well.) It is an elastic deformation, if an unusual one. Material strength does not affect buckling any more than it affects bending or torsional deflections. Like bending and torsion, material modulus does affect buckling, as does geometry.

Mechanical engineers typically study column buckling, but plates, beams, shafts, arches, helical compression springs, and shells also buckle if too slender (Figure 13-1). Perhaps you can analyze for buckling, and you surely can test for it, but you must remember to do it. Whenever you visualize and plan a load path, consider buckling throughout the design.

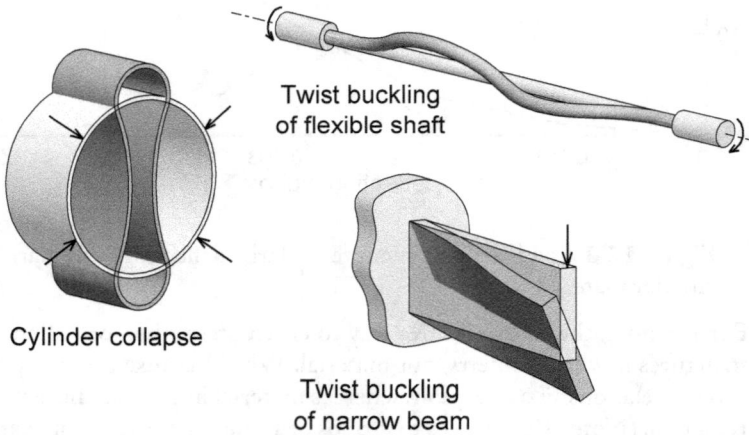

Twist buckling
of flexible shaft

Cylinder collapse

Twist buckling
of narrow beam

Figure 13-1 Buckling occurs not only in columns, but also in many other slender structures.

If you suspect buckling, analyze or mitigate. As always, improving geometry will be the best choice for preventing buckling. Stiffen long struts or webs with ribs and three-dimensional triangulation. Also consider supports analogous to mid-length guy-wires—a little lateral support will go a long way.

Do not overlook the advantages of buckling either. Remember, it is not fundamentally a failure, but is an elastic deformation with unusual characteristics. You can use the unstable nature of buckling like a load switch: no deformation up to a switching load, an avalanche of deformation at that load, a return, unharmed, to the start position with load removed. Some snap-acting switches use this principle by buckling a column or a spring.

14. Analyze and test for trends and relationships.

In design, the value of engineering analysis often is not in correctly answering an analytical problem, but in gaining insight into the behavior of a device or system. This insight will lead you to better designs faster.

For example, consider the scissors jack shown in Figure 14-1 . A basic force balance is:

$$Fw = Fa / \tan \phi \qquad\qquad (Equation\ 14\text{-}1)$$

where Fw is the resultant force, Fa the applied force, and ϕ the angle of the connecting bars.

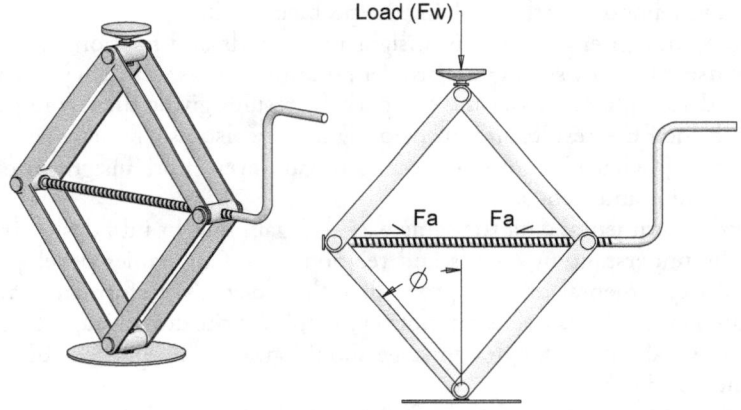

Figure 14-1 Scissors jack. Knowing relationships between parameters is a useful guide for design.

Knowing this force relationship will help you immensely to design this device. The jack converts handle input to lifting throughout its range, but with remarkably different performance over this range. As the jack extents upward, the angle ϕ gets ever smaller as does the tangent of the angle. With the tangent in the denominator, the resultant force becomes quite large for small angles. On the other hand, if the angle is less than 45°, the lifting force is less than the applied force.

A motion analysis tells you that the rate of height increase is rather modest when the force relationship is advantageous, and a structural analysis tells you if the parts are strong enough. (Don't forget buckling!)

If you did not know the force and motion relationship, you would not know where in the operating range the jack best matches your requirements—you would not even know there is a difference, at least not until you tested it. But by knowing the force relationship, you understand the trend:

the smaller the angle, the better the lifting force transfer—and dispropor-
tionately so.

So even if you don't need to, don't want to, or can't analyze a system for
an absolute answer, develop or look up the relationships among parameters
so you understand the trends in a device's behavior.

Engineering analysis software such as finite element analysis (FEA)
predicts mechanical behavior even on complicated systems and geometry.
Nevertheless, an analysis alone offers limited insight into the relationships
and trends. Unlike strength of materials or elasticity theory analyses, no ex-
plicit relationships are stated. This is unfortunate because—restating the
theme of this section—understanding the relationships among variables is
often more important than solving the problem itself.

Many design engineers gain insight into trends and relationships even
when using FEA by solving the problem traditionally as well. They calculate
by hand a simplified model and compare the results, giving them confidence
in both the FEA results and in the design. They also use "families" of FEA
results and "what if" computer results to gain even more insight into the
relationships and trends.

Use experimental data the same way—to gain insight into design problems by understanding trends and relationships. Companies develop sta-
tistical experimental data to prove that their device's performance meets
requirements. This is noble, but not very helpful at the design stage. Instead,
experimental data at the design stage should give you insight into both robustness and behavior.

Finally, recognize the power of relative and normalized data. We need a
reference to make sense of most things, so your analyses and test data are
more useful if they compare conditions or relate to a mean.

15. Identify contingency plans to minimize risks in design.

Product quality assurance programs manage risk using failure modes and
effects analysis, fault tree analysis, design validation testing, and so on. These
are good but inadequate safeguards against design errors. However useful,
they only identify problems, often rather late in a project, and do not help to
fix them. That usually comes back to design.

Include as part of your design process a review for anything that could go
awry; anticipate possible failures and identify corrections should they come
about. Always give yourself an out. Do not limit your definition of failure in
this case to hazards, but include function, robustness, production, service,
and life.

Leave a little extra space. Consider alternative materials. Investigate more expensive components. Ask yourself questions like:

+ If three bolts are not strong enough (because, for example, you are as yet unsure of the loads), can I simply increase their size? Change to four? Or must I drastically redesign parts to accommodate bigger bolts?
+ If our life testing shows that this linear slide wears out too quickly, can I add a rolling element linear bearing without major redesign? Will that unquestionably solve the wear problem? Can we accept the increased cost?
+ Can I easily add ribs in this molded part if the bosses are not stiff enough?
+ If this plating proves to be inadequate for corrosion protection, can I change to stainless steel? Or would I face a major redesign if I change materials?
+ If the vendor supplies components slightly out of specification (because, for example, they use a new process), will the design still work? If not, what will I have to do to use the parts?
+ If we find that users break this device more often than we (or they) would like, can we service them easily? (That is, assuming prevention is impossible!)

PART III

SOME
PRACTICAL
ADVICE

---

16. Avoid press fits.

Perhaps because the Machinery's Handbook [45] has tables for press fits, we are lured into using them. Besides, the geometry is simple, and parts count is minimum. Rather standard, accepted practice, right?

No! Why? Because a press fit is overconstrained. Press fits require tight tolerances, generate uncontrolled friction, and create assembly stress, all of which good designers try to avoid. Furthermore, they are hard to assemble, and disassembly might be impossible.

Think press fits, and most of us think of dowel pins pressing into one or more mating parts. But press fits tempt designers in other geometries as well, all of which are sub-standard design.

So what should you do? Some alternatives to press fits are:

+ Elastic fits
+ Snap fits
+ Tapered fits

Elastic fits are the "elastic constraint design" version of a press fit (Figure 16-1). If your manufacturing process allows it, build in some elasticity to accommodate the required deflection.

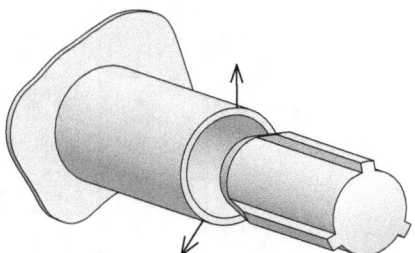

Post ribs flex cylinder out-of-round.

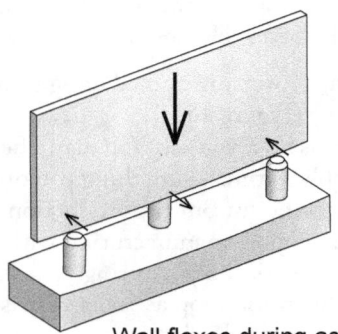

Wall flexes during assembly.

Figure 16-1 Elastic fits replace overconstrained press fits.

In the first example shown, the traditional loading of a press fit, radial expansion, is replaced by cylinder wall bending. (This is an example of where you want to use bending!) Tight tolerances are no longer required. The holding force will be limited by the elasticity, but then again, it will be much better controlled. The second example uses a flexing wall for the elastic fit.

Snap fits are the "functionally independent" equivalent of press fits. Separate retaining and locating functions, and you eliminate tight tolerances (Figure 16-2).

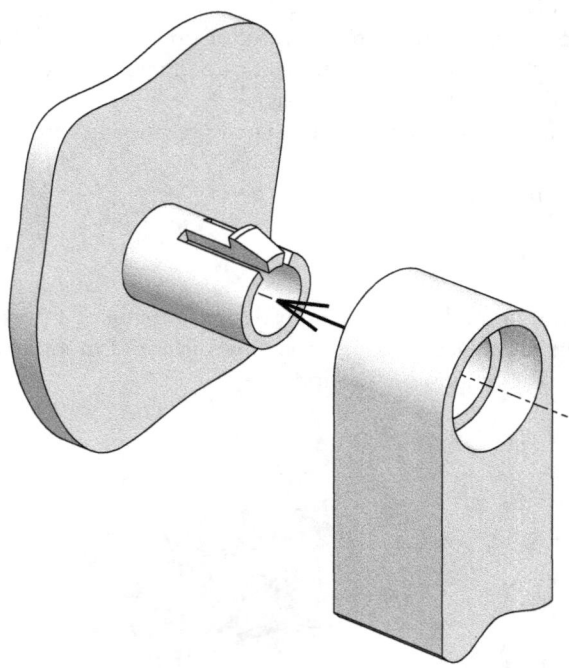

Figure 16-2 Snap fit post avoids the over-constraint of a press fit, and relaxes tight diameter tolerances.

Tapered fits are still acutely overconstrained designs, but you will have a lot less trouble with assembly (Figure 16-3).

"So I should never use press fits?" you ask. You may. They are plenty handy for one-ups, test and assembly fixtures—anything you or the machine shop can tinker with or correct on-the-fly. But for production items, it is best to avoid press fits. Or plan to support manufacturing with design validation, substantial quality control, and assembly fixturing.

As a final note, a press fit in thermoplastics is worrisome, substandard design. Replace it with snap fits or elastic fits—or risk creep and fractured parts.

Figure 16-3 Tapered pin improves assembly, but does not eliminate over-constraint.

17. Use closed sections or three-dimensional bracing for torsional rigidity.

Most of us only rarely encounter challenging problems in torsional stiffness. The shaft or structure is probably rigid enough in torsion when we design the other characteristics of the system. (That is another thing not to trust!) But for a problem with torsional rigidity, a little insight goes a long way.

First, closed sections are overwhelmingly stiffer than open sections in torsion.

Traditional elasticity theory offers the membrane analogy for torsional bars. This analogy says that the polar moment of inertia is proportional to the volume contained in an inflated membrane attached to the edges of the section. This visually shows why closed sections, even of thin-walled tubes, are far more rigid than the same, non-closed section (Figure 17-1).

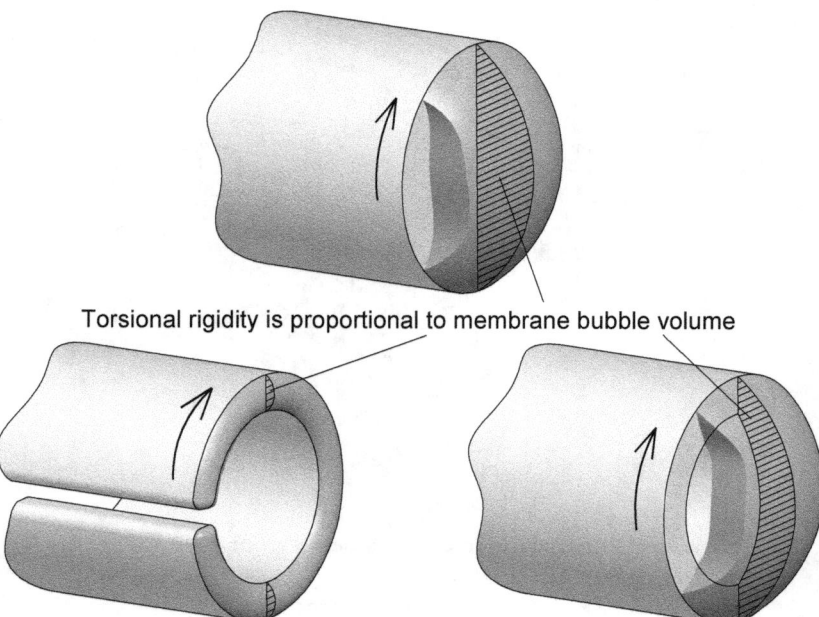

Torsional rigidity is proportional to membrane bubble volume

Figure 17-1 Membrane analogy for torsion of solid bar, thin-walled tube, and split tube. The relative torsional rigidity is proportional to the inflated membranes' volumes.

"OK, I get it," you say, "use closed sections. What's the big deal?" But what do you do if you cannot close the section? This is a common problem for cast, molded, and extruded forms. Play with the 2D section form of an open-section torsional bar all you want. Add ribs, make an I-beam, X-form, U-shape, even make it larger. Nothing helps much to increase torsional stiffness in open sections. The membrane analogy predicts this and helps you visualize it.

But there is a solution, one that the membrane analogy does not suggest: add three-dimensional bracing. That is, triangulate the structure using the third dimension (Figure 17-2).

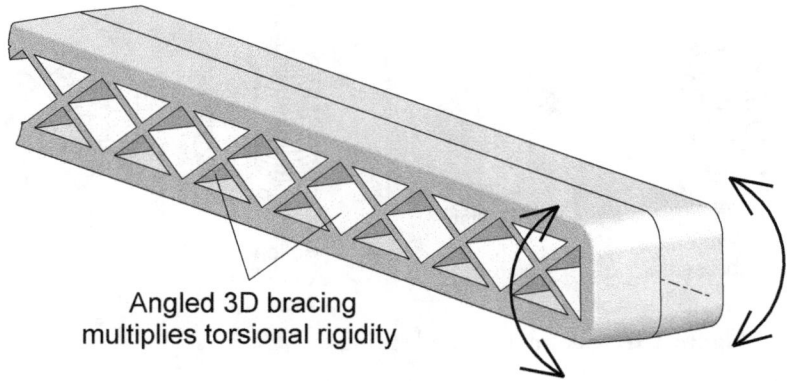

Angled 3D bracing
multiplies torsional rigidity

Figure 17-2 Three-dimensional bracing of open section for tor-sional rigidity. This is especially useful in cast and molded parts.

18. When designing springs, use a low spring rate and a high initial deflection.

Machine design textbooks and engineering handbooks offer widely available, and quite sufficient material for analyzing springs. You can find plenty of handy spring design software, which I find invaluable, and which everybody designing wound springs these days uses, to simplify the task. Of course, there are plenty of watch-outs: stability, fatigue, corrosion, tolerances, natural frequencies, all of which you studied if you have studied springs at all.

What all of this does not tell you is how to select the operating point of a spring, but I recommend designing springs with the lowest possible spring rate, and operating them at the largest possible deflection.

But why, you ask, would I want to operate a spring near its maximum allowable stress? Won't that precipitate failure? No, at least not when you follow the customary design guidelines as well.

Springs exert force and store energy. The concept presented here applies to any spring, but wire wound springs are the most common. Figure 18-1 shows useful types of wound springs.

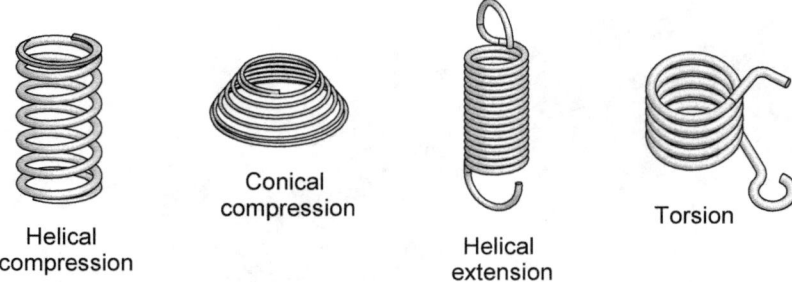

Conical
compression

Helical
compression

Helical
extension

Torsion

Figure 18-1 Common wound springs.

Helical compression springs are probably the most common type, this because they are inexpensive, easy to assembly, and robust. Use them whenever possible. Conical springs can be made to "telescope," reaching as little as one wire diameter in solid height, their main advantage. Extension and torsion springs need formed ends, are often difficult to assemble, and are less robust. Invaluable for their function in some applications, they are still second choices after compression springs.

Compression and extension springs are, perhaps surprisingly, torsionally-loaded beams. The maximum force a wound compression or tension spring creates is dependent on the material strength, the coil diameter, and the wire diameter. Free length, number of coils, shear modulus of elasticity, and end configuration affect the spring rate and the operating points, but not the maximum possible force.

Perhaps also surprising, torsional springs are flexural beams. The maximum force a torsional spring creates is proportional only to the material strength, and the wire size. The coil diameter and number of coils affect only the spring rate and the operating points (because these control the beam length), but not the maximum possible force.

Figure 18-2 shows a load-deflection diagram for a helical compression spring. Think of designing springs as selecting the correct load-deflection diagram together with the correct operating points and range on that diagram.

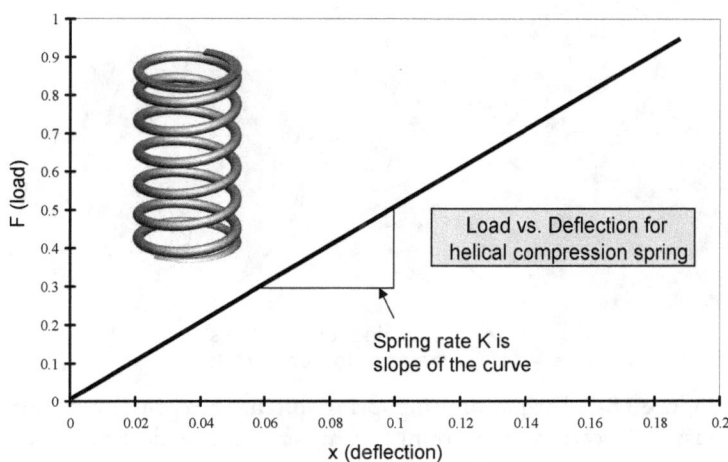

x (deflection)

Figure 18-2 Load-deflection diagram for a helical compression spring.

The diagram shows this spring's curve as a straight line, meaning it follows Hooke's law:

$$F = -K \times x \qquad \text{(Equation 18-1)}$$

where F is spring force, x is deflection, and K is the spring constant.

Designing coil springs is unavoidably trial-and-error. You must assume a geometry, calculate the result, adjust the geometry, and recalculate again until you are satisfied with the result.

Force is the most important parameter in designing springs. Usually, this force must be exerted at a particular position, and this combination of "force at a height" is the operating point. Sometimes, rather than a single point, there is an operating range or multiple points, but in all cases, the spring operates at or below, around, or above an operating point. For example, a torque-limiting spring clutch uses spring force to control the slip torque (Figure 18-3). The operating point is the spring's length when a ball fully nests in its detent hole, and the range is defined by the position when the ball is fully disengaged.

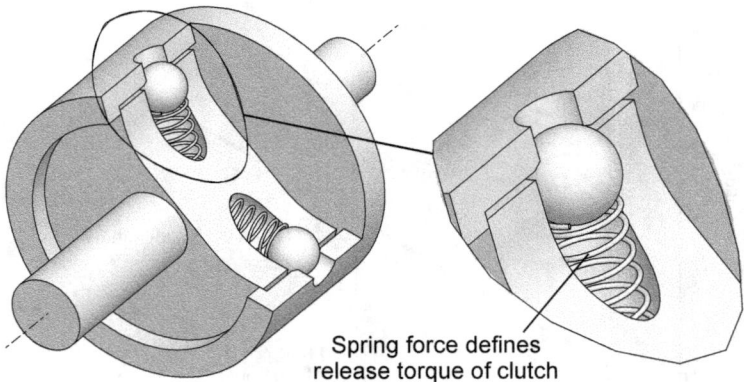

Spring force defines
release torque of clutch

Figure 18-3 Torque-limiting spring clutch. The spring's rate, not just its force, is an important parameter in spring design. Lower spring rates are usually better.

In almost all cases of simple spring design, it is desirable for the force to remain as close as possible to the force at the nominal operating point throughout the operating range. The spring clutch of Figure 18-3 works because the spring force remains close to the initial force as it deflects; that is why we use a spring. We do not want the spring force to be many times higher in the disengaged position, but instead want the spring force to remain near the engagement force through the disengagement range. The same is true for almost all spring uses.

The previous paragraph implies that the lowest possible spring rate is the most desirable. The lower the spring rate, the less the force will change for a given deflection. It also implies a relatively large initial deflection, the deflection to get to the nominal operating point. Perhaps surprising then, good spring design often means increasing the stress! In fact, many optimum spring designs operate near the maximum allowable stress. So, repeating, within the bounds of all other constraints, the optimum design (usually) is that with the lowest spring rate and highest initial deflection; this will give you the best functional behavior.

But there is more. Equally important reasons for a low spring rate and high initial deflection are manufacturability and robustness. Assembly parts' tolerances, spring dimension tolerances, and force adjustments are all more forgiving with lower spring rates. This is because for a given dimensional variation, the force varies less with lower spring rates than with higher spring rates (Figure 18-4, FL vs. FH, respectively). In other words, the flatter the curve, the better. Thus, for the spring clutch of Figure 18-3, the variation in release torque due to part tolerances is smaller with a lower spring rate, improving manufacturability. Moreover, wear and corrosion, for example, affect release torque less, improving robustness.

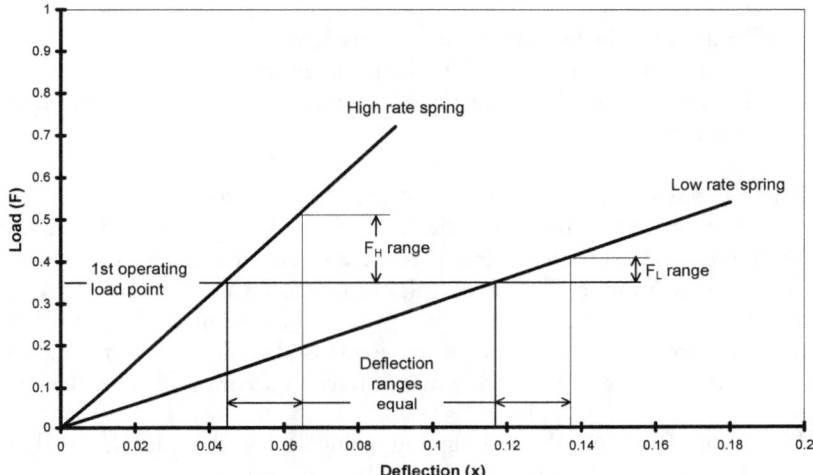

Figure 18-4 Force range vs. deflection range for different rate springs. Lower spring rates mean less force variation over the operating range.

Yes, but won't a high-stress spring have a much lower life in cyclic applications? Probably. But this is not as bad as it first seems. Fatigue failure models all account for both mean stress and alternating stress. A low-rate, high-deflection spring has a higher mean stress compared to a high rate spring, but this is offset by a lower alternating stress. In any case, you will still want to use the lowest-rate, highest-deflection design that meets the fatigue life.

So a well-designed wire-wound compression or extension spring is likely to have (relatively-speaking!) a long free length, thin wire, and many coils to give it a low spring rate. Its initial deflection will be large, its operating force is nonetheless small, and its change in force over the operating range is minimal. Spring equations reveal the following mathematically, but I like to keep these trends in mind:

Spring Rate
+ To decrease spring rate, increase the number of coils.
+ To decrease spring rate, decrease the wire size.
+ To decrease the spring rate, increase the coil diameter.
Note: The following comments about stress and load are for a given deflection.
Stress
+ To decrease the stress, decrease the wire size (this is perhaps counterintuitive!).
+ To decrease the stress, increase the mean diameter.
+ To decrease the stress, increase the number of coils.

Load
+ To increase the load, increase the wire size.
+ To increase the load, decrease the number of coils.
+ To increase the load, decrease the mean diameter (also perhaps counterintuitive!).

There is one consistent exception to the aforementioned philosophy of low spring rates and high initial deflections. Some springs must not remain in a stressed position for long time periods, and plastic springs of any form are a notorious example. They stress relax rapidly, even at room temperature, and will cease to be springs sooner than you will like. Any application that requires no pre-deflection (i.e. constant load) is also likely to come with a higher relative spring rate, which you will have to accommodate in the design.

Here is one final note about designing springs. Savvy designers know that the maximum force of a spring is limited by the amount of material in the spring. You can play with spring parameters all you want, but if you need more force, you need more material. This becomes painfully clear if the space for a spring is already fixed, but you need more force. Savvy designers leave some extra space just in case, or test early and thoroughly. Some do both.

19. Minimize and localize the tolerance path in parts and assemblies.

You will at some point need to dimension and tolerance parts in your designs. If you include this as a part of design itself, you will improve your designs' manufacturability. You should dimension based on function, and you should design for locally-controlled critical dimensions.

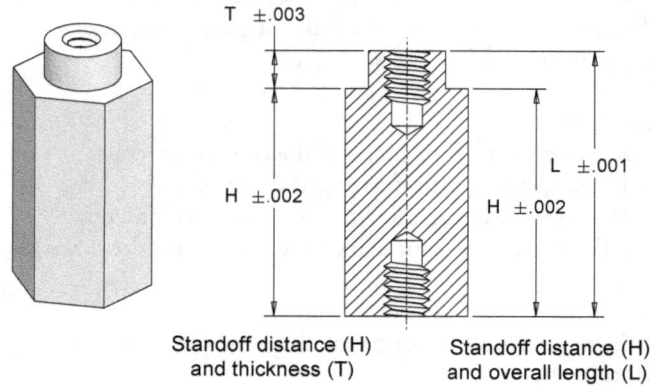

Figure 19-1 Two dimensioning schemes for a standoff.

Figure 19-1 shows two possible dimensioning schemes for a standoff as might be used, for example, to mount an electronic printed circuit board at a distance from a housing floor.

On the left, there is a dimension H with a tolerance that defines the height off the floor and a dimension T with a tolerance that defines the captured board space. An alternate dimensioning scheme is on the right. The second dimensioning scheme requires a much more accurate overall height. For this example, dimensioning the part based on its function has given considerably more forgiving manufacturing tolerances: T ±.003" for the left scheme vs. L ± .001" for that on the right, giving an overall tolerance to the length L of ±.005" vs. ±.001".

The advantage comes from minimizing the tolerance path or tolerance stack-up, which is not at all a part of tolerance analysis. It is a part of design.

Sometimes you can do even better by designing the parts and assembly so that important dimensions are easiest to hold. Figure 19-2 shows two versions of an O-ring seal joint, circumferential and face.

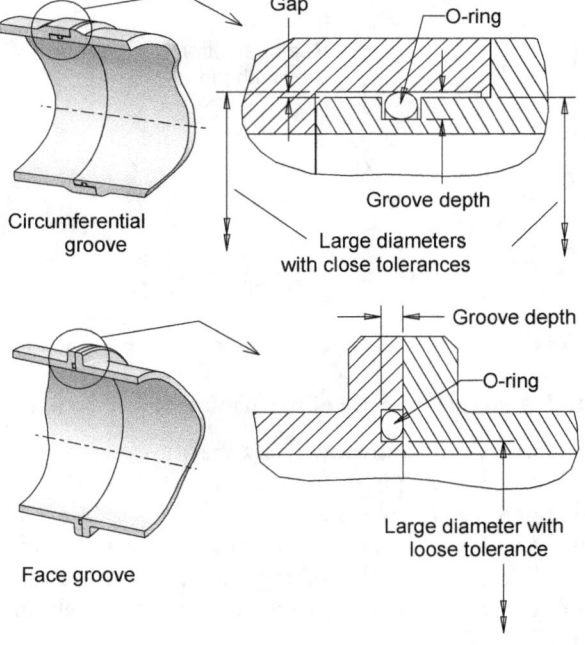

Figure 19-2 O-ring seal joint comparing circumferential with face seal dimensioning. The face seal's important dimensions are probably easier to manufacture.

The circumferential version requires gap and groove to be held to close tolerances. The gap and groove dimensions are governed by both mating parts' diameter dimensions. To hold these diameters to the required close tolerances could be quite a manufacturing challenge, especially for a large diameter pipe.

For the face seal, a single dimension governs the groove depth. The gap tolerance problem disappears because the parts' flat faces seat together during assembly. (Manufacturing flat faces is rarely much of a challenge.) Thus, the face seal design localizes the required dimensional tolerances to a single part and to relative, rather than absolute, dimensions within that part.

20. Use mechanical amplification to reduce failures.

An amplifier is, by one definition, any device that uses a small amount of something to control a larger amount. This can apply to mechanical design. The simplest of examples is using a cotter key as a retainer only, rather than to carry load (Figure 20-1).

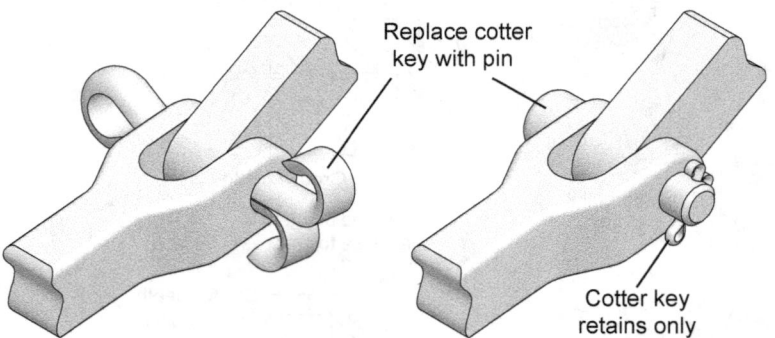

Figure 20-1 A simple example of mechanical amplification.

OK, I said this several times already. This a specific example of functional independence, or one of designing the load path. Perhaps these would be obvious enough, but keep this obvious example in mind. Use load-bearing parts or structures to bear loads, and use latches, fasteners, and adhesives to hold components in place. A small latching, holding, or adhesive force can control a much larger structural force, which is often quite helpful for user-centered design (Figure 20-2).

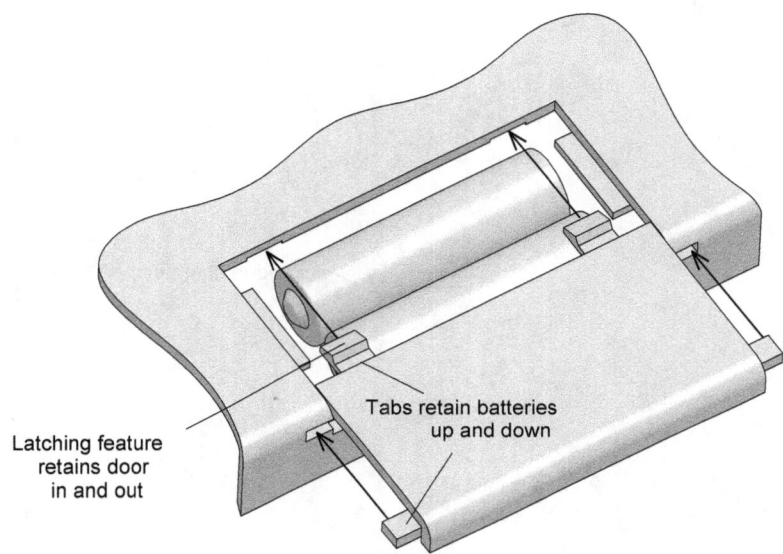

Figure 20-2 Battery door with mechanical amplification. The door's latching features do not retain the batteries.

21. Include lead-ins in assembled designs.

A key feature for both manufacturing floor and user-centered design is the lead-in or chamfer. If you have studied design for assembly, you know how important chamfers are in design for assembly. But do not overlook their value in user-centered design, or anywhere else.

The tapered end of a dowel pin (Figure 21-1) provides three important functions:

1. Starting, or spearing, the parts together
2. Aligning to the desired final position
3. Directing the applied force

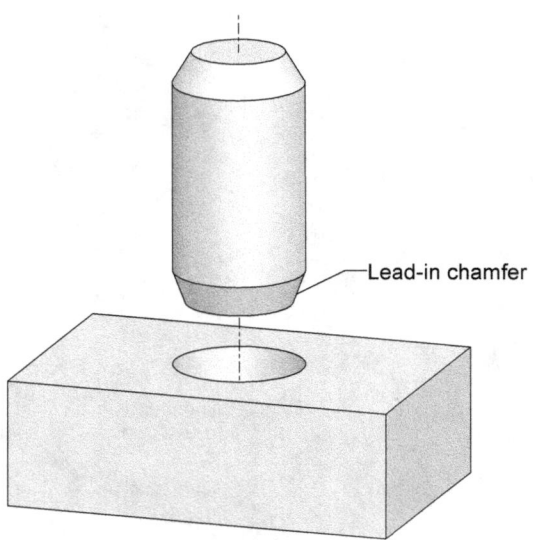

Figure 21-1 Dowel pin with lead-in.

Without a lead-in, the dowel pin would be impossible to spear into the hole and almost impossible to align perfectly with the hole. Further, it would be impossible to press the pin into the hole without damaging the hole. Chamfering both hole and pin eases assembly even more.

Everything about changing a flat tire on an automobile is detestable, but my least favorite part is remounting the wheel onto the lugs. Larger lead-ins here would make a difficult task remarkably easier. Unfortunately, lead-ins require space, which is probably why there are no lead-ins on automobile wheels. Because of this space requirement, it is important to plan for lead-ins during design. Overlooking them may become costly if they must be added later but there is no space.

Lead-outs are the opposite of lead-ins, and are used in items that require easy disassembly. An example of a lead-out is the cover of a food storage container. As in this example, lead-ins and lead-outs often are dissimilar to give a differential feel: easy to close, harder to open.

22. Design assemblies to be self-locating, self-fixturing, self-securing, self-aligning, self-adjusting.

Whenever possible, try to design location, adjustment, fastening, and alignment as intrinsic characteristics of component parts rather than relying on assemblers, users, or fixtures to do the job. The parts themselves will usually do it cheaper and more reliably.

Any time a device includes an assembly process, whether for the manufacturing floor or for user-centered design, the assembly should be self-locating or self-fixturing. Help assemblers, users, or yourself by providing clear cues and features (Figure 22-1).

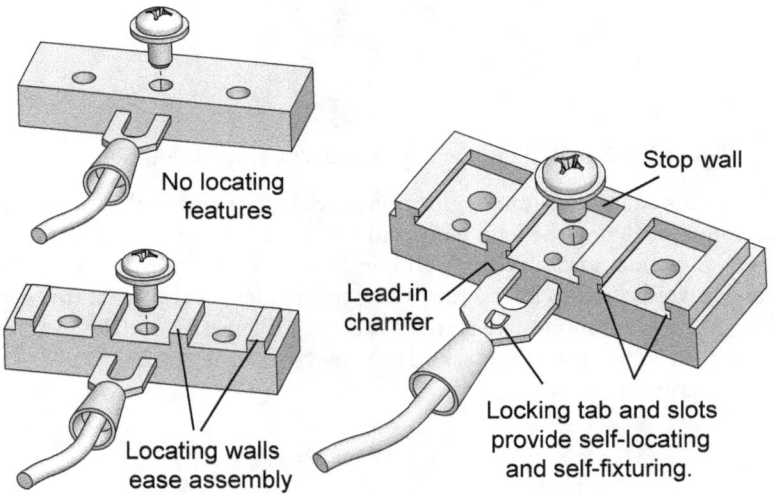

Figure 22-1 Self-locating and self-fixturing examples of a spade terminal.

A more sophisticated concept is self-alignment as in the disk brake of Figure 22-2. The caliper assembly is free to float back and forth so that it can align itself perfectly with the disk. This self-aligning capability is important not only for ease of manufacturing but also for accommodating wear.

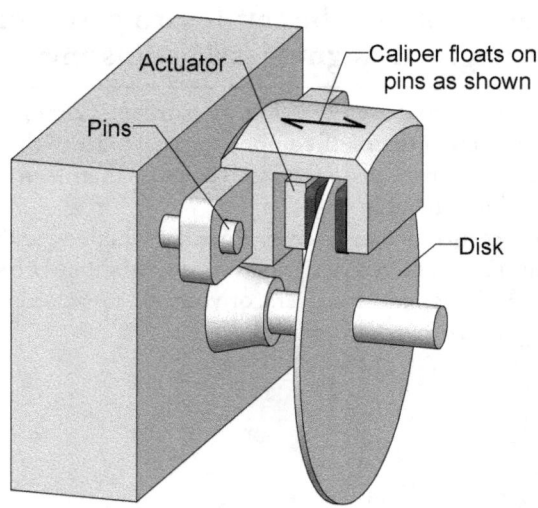

Figure 22-2 Self-aligning caliper of disk brake assembly.

I like this example because, along with self-alignment, it illustrates exact constraint, localizing the load path, and functional independence as well.

Self-adjustment is found, for example, in some brakes and clutches. Many cable-actuated clutches have a ratcheting mechanism at the pedal that repositions itself to accommodate clutch plate wear (Figure 22-3). The pedal and cable return to home positions, and the pawl grabs the ratchet regardless of parts variations or wear.

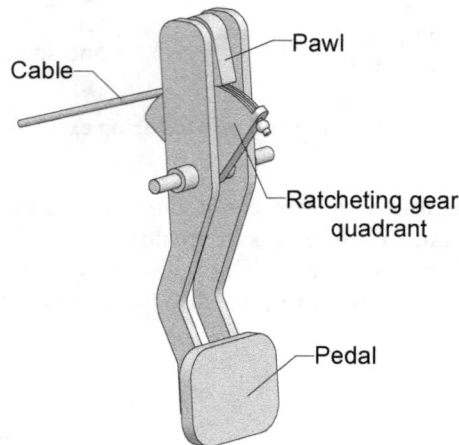

Figure 22-3 Self-adjusting clutch cable assembly.

Of course, design is a compromise, so you will not want to strictly adhere to all of these assembly design rules. Including these features may cost more than they are worth, but know that before including or rejecting.

23. Use self-assembling symmetry to create a whole from two halves.

Sometimes it is impossible to create all the functional features required in a single part at a low enough cost. This often happens with undercuts in molded, cast, and even machined parts. An example is an internal flange in a cylindrical part (Figure 23-1). The only acceptable alternative may be to split the design into two parts, then assemble them together.

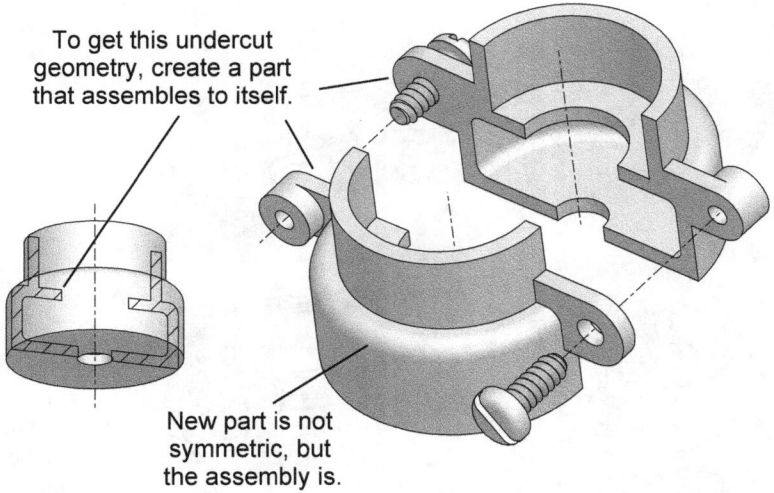

To get this undercut geometry, create a part that assembles to itself.

New part is not symmetric, but the assembly is.

Figure 23-1 A cylindrical assembly with an undercut. Self-assembling symmetry allows the parts to be identical.

There are several solutions, of course, but splitting the part as shown allows two of the same part to be assembled together, forming a whole. This solution is sometimes best, especially to minimize capital costs, because only one part is manufactured. It is particularly useful for molded and cast parts.

I call this self-assembling symmetry. Perhaps not a mathematically rigorous definition, it is utilitarian nonetheless. (Symmetry's accepted definition is that a shape is symmetric if it looks exactly the same after being transformed in some way—reflected, rotated, slid, expanded, contracted. [52]).

The part is not symmetric, but the assembly is, 180 degree rotational symmetry in the above example's case. This assembled symmetry is itself the key to designing features.

Although the part itself has no symmetry of its own, it has the unusual characteristic of features working together in pairs. With self-assembling symmetry, after a transformation, typically 180 degree rotation, the object will assemble to itself to create a new, useful object.

This is especially advantageous with large housings, for example, whose tooling is quite expensive (Figure 23-2). Rather than making two different molds or mold cavities, a single mold at half the cost can be made. Higher part cost, if any, is justified by lower capital costs, higher production volume, and fewer items to inventory. Then there is the obvious assembly advantage of handling two of the same rather than two different parts, a dream for assemblers!

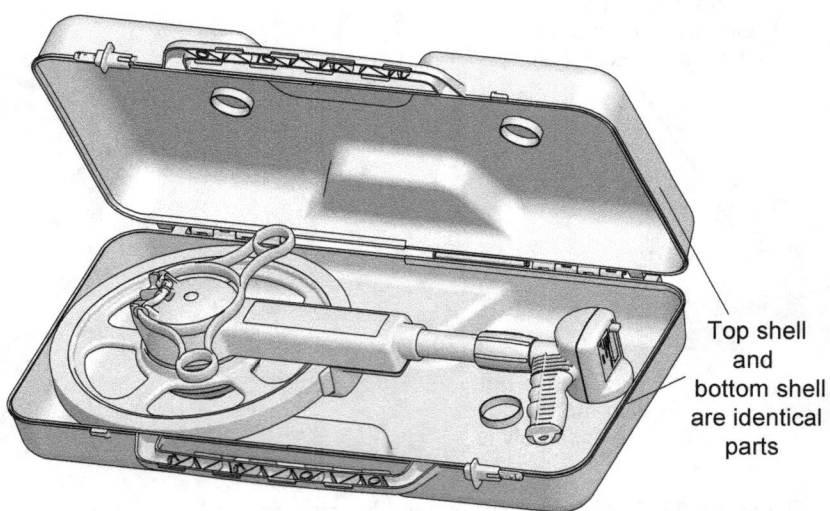

Top shell and bottom shell are identical parts

Figure 23-2 Identical large housing halves. Identical halves can reduce tooling costs and simplify assembly. (Used with permission of Meter-Man, Inc.)

Although challenging, integral hinges, latches, handles, snap fits, and the like can all have self-assembling symmetry. 3D solid modeling is especially helpful to work out the required geometry. You begin with the basic half shell or half part, assemble it to itself, then start adding mating features like a latch or a hinge. The 3D model lets you visualize quite nicely as you go how to configure the required features (Figure 23-3). (This designer would never attempt creating self-assembling symmetry parts without 3D solid modeling!)

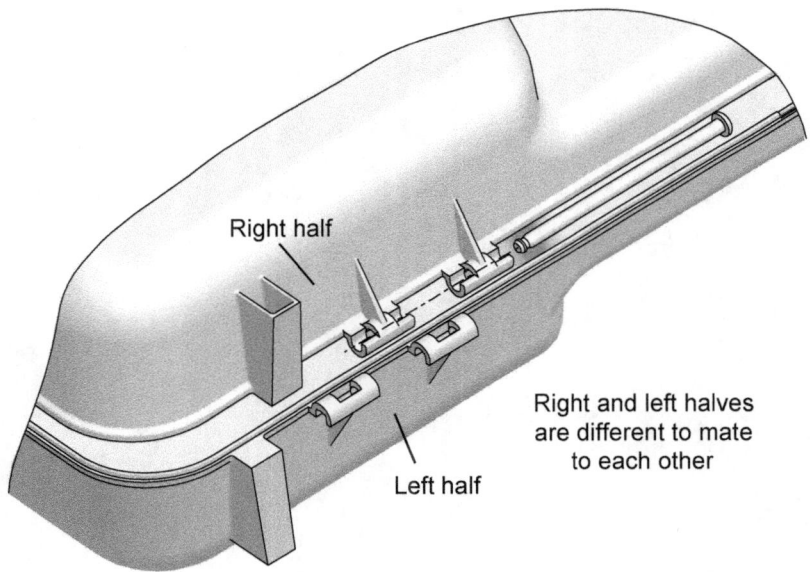

Figure 23-3 Feature detail of self-assembling symmetry. (Used with permission of Meter-Man, Inc.)

Appendix A: Rules for Exact Constraint

Although too abstract for the main body of this book, geometry-based rules for exact constraint can be useful, even if only for gaining insight into the subject. Here are some, along with my explanations. (Reprinted with permission from Eastman Kodak Company [9].)

+ No two constraints are collinear

This means that the nesting forces for any two constraints must not be along the same line. See Figure 3-10, #5 for an example.

+ No four constraints are in a single plane

Since there are only three constraints required or possible in one plane, four would be overconstrained. See Figure 3-10, #6 for an example.

+ No three constraints are parallel

This is analogous to saying that two points determine a line. Three points is an over-constraint.

+ No three constraints intersect at a point

This is a variation of the previous statement, but for a curve rather than a line. See Figure 3-10, #8 for an example.

We can add the following for 3D space:
+ No four constraints are parallel

This is analogous to saying that three points determine a plane. A fourth point would produce over-constraint.

+ No four constraints intersect at a point

This is a variation of the previous rule, but for curved surfaces. An example is the trihedral receptacle and ball used in Figure 3-5. A four-sided receptacle would be overconstrained.

+ No four constraints are in the same plane

In effect, this rule says that a rotational constraint cannot go through the axis of rotation. This is logical since such a constraint would have no moment arm to prevent the rotation.

APPENDIX B: NESTING FORCE WINDOWS IN EXACT CONSTRAINT DESIGN

To construct a two-dimensional nesting force window (adapted from Blanding [12]:

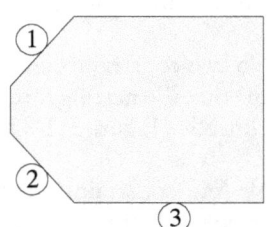

1. Label the constraint contact points 1, 2, and 3.

2. Draw lines through the three nesting force vectors to intersect each other. (Perpendicular to the tangent line at the contact point.)

3. Label the intersections by convention 1-2, 2-3, and 3-1. These are the instant centers of the constraint system.

4. Imagine pinning the first instant center, 1-2, then determine which rotational direction about that center will force contact with constraint 3. Draw an arc arrow at the instant center and label it CW (for clockwise) or CCW (for counter-clockwise).

5. Do the same for instant center 2-3 with constraint 1, and instant center 3-1 with constraint 2.

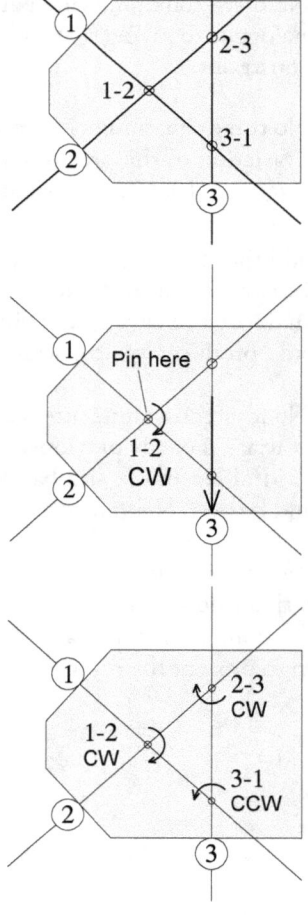

6. Connect with a bold line any instant centers for which the direction is the same (CW with CW, CCW with CCW).

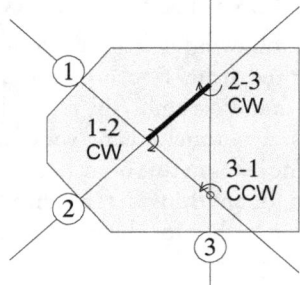

7. If the bold lines form a triangle, the nesting force window is outside this triangle. (But beware of direction, for it circles the triangle!) If not, draw a bold V outward from the instant center *opposite* the first bold line on the two constraint lines.

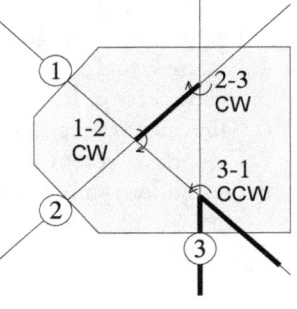

8. Draw bold lines outward from the instant centers at the ends of the first bold line, in line with the second and third bold lines. The nesting force window is complete.

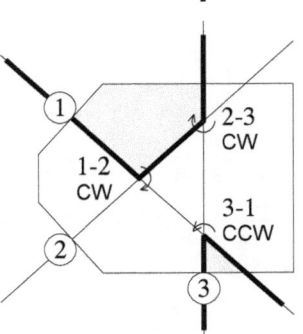

9. For two parallel constraints, the procedure is conceptually the same, but one instant center is at infinity.

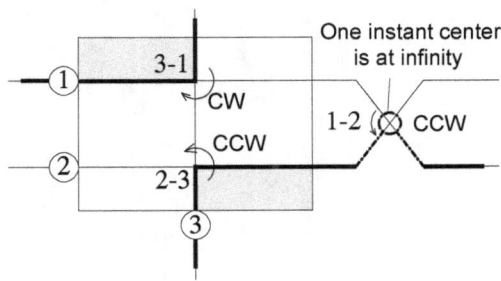

APPENDIX C: DESIGN FOR ASSEMBLY RULES

The following list gives design recommendations for ease of assembly (adapted from Boothroyd and Dewhurst [49]). It was originally written with hand assembly in mind, but it applies equally well, with some additions, to automated and robot assembly.

The list is in this book because these are good rules for all design. Consider them for anything that requires assembly, whether you will make one or millions, whether for the manufacturing floor or an end user.

1. Reduce part count and part types
2. Eliminate adjustments
3. Design for self-alignment and self-location
4. Ensure visual and physical access
5. Ensure ease of handling bulk parts
6. Minimize reorientations during assembly
7. Design to prevent improper assembly
8. Design for symmetry or obvious asymmetry

APPENDIX D: WITH EXPERIENCE COMES WISDOM

The experienced designers interviewed for this book offered valuable insight into mechanical design. I incorporated many of their comments into *The Elements of Mechanical Design*. Here are some samples:

Have a back-up plan! Ask yourself where the risk areas are.

Start with the simplest principle and then add complexity as you need to.

Really creative people can take unrelated things and put them together in ways that someone else isn't normally going to think of.

The best designers are hands on.

Modeling something full scale is so illuminating. When you hand over a model to a client, it's a revelation—either a good one or a bad one.

Prototyping really helps us communicate with non-technical people. They see it on the computer screen—the shape and everything—but they don't get an idea of the size.

I just did something recently like that. When it was actually done, it seemed a LOT smaller than I expected.

For complex assemblies I found that it is well worth making engineering models of small parts of the assembly and test them…simplified models… principle models.

The notion that you can … catch everything in CAD or FEA is nonsense.

If you did a study of number of iterations versus (quality) of the design, you'd find that there is a direct relationship.

We make a scale model out of mat board and hot melt glue, and then we stress it. Then we say, "Oh look, it buckles over here!" It's so much quicker than analyzing something with 20 parts in it. Especially for 3D structures it's fantastic.

It's funny how wrong people are about stiffening something.

In machine design, you're more often designing for stiffness than for strength.

Stiffness is like a springy thing. Strength is more where it fails.

I've never had much luck with predicting the effects of friction.

Typically, mechanical design consists of transmitting or converting motion, or devising joints that hold it all together.

By experience you learn that almost every linear motion is more difficult to do than rotary motion.

One example (of bad design) is linear sliding devices. Engineers forget about the bureau drawer effect.

The whole linear binding thing is a major, major difficulty.

In a lot of the stuff we design…there's a lot of room for common sense and cleverness, but not so much need for … analysis to get the answer.

Sidestep the need for a detailed analysis.

If you have a requirement you don't think you can meet, instead of trying, you should design around it. If you can't solve the requirement, just get rid of it.

When you find a problem, fix the cause, not the effect.

I hate press fits. They're so sensitive to tolerances. I've had them fall out and kill my design.

If this is where you want the load, and this is where it's applied, put them together and have nothing in between, and then you'll have nothing to fail.

If you can make a measurement either differential or ratiometric, it eliminates a lot of errors.

Cross leaf flexures are interesting things that most people don't know about.

I still will do a layout, but it will be usually in solid models. You do a solid model analogous to the old 2D layouts, but it can be very block-like…not a lot of detail.

I do a little bit of sketching. But I try to get to CAD modeling as soon as possible so I can see it and understand it.

One thing I have seen is a tendency to try to get too detailed too quickly.

If anything is a little frustrating, it is that engineers (do not) see the big picture of things.

Industrial designers come up with many ideas first, then pick the best or the best combination. Whereas mechanical engineers simply go down the first obvious path and add afterthoughts to address issues that come up.

When you're an inexperienced designer, you'll think of a bunch of ways of solving a problem, and not know which one of them is good. You don't really have much framework in which to evaluate your ideas.

If you nurture a few ideas along from the beginning, you're more impartial… and you might just change your mind, too.

Things can be simple and complex at the same time.

It's very common to make things more complicated than they should be.

The design is not done when you can't add anything more to it; the design is done when you can't take anything else away.

TEXT NOTES AND REFERENCES

1. Most design literature describes simplicity, in one form or another, as a fundamental concept [1,2]. It is, for example, the second of Suh's two axioms. (Minimize the information content of the design.) In *The Elements of Mechanical Design*, simplicity is directly promoted, without abstraction.

This discussion comes dangerously close to advocating simplicity in, for example, consumer products. *The Elements of Mechanical Design* is about design at the component and assembly level. Quantity and complexity of product features is something different altogether, governed by a different, larger process.

References:
1   Nam P. Suh, *The Principles of Design*, Oxford University Press, 1990.
2   G. Pahl and W. Beitz, *Engineering Design, A Systematic Approach, Second Edition*, London: Springer-Verlag, 1996.

2. Techniques for functional decomposition are found in design process books [3,4]. Functional independence has been proposed as one of two fundamental axioms (along with minimum information) applicable to design by Suh [5]. In Suh's words, "Maintain the independence of functional requirements." Other authors promote similar concepts, e.g. Clarity [6]. In *The Elements of Mechanical Design*, the same idea is stated as directly as possible, without abstraction.

References:
3   David G. Ullman, *The Mechanical Design Process, Third Edition*, McGraw-Hill, 2003.
4   Karl T. Ulrich, and Steven D. Eppinger, *Product Design and Development*, McGraw-Hill, 1995.
5   Nam P. Suh, *The Principles of Design*, Oxford University Press, 1990.
6   G. Pahl, and W. Beitz, *Engineering Design, A Systematic Approach, Second Edition*, London: Springer-Verlag, 1996.

3. Exact constraint has suffered obscure and fleeting mention in science and engineering throughout the history of technology. Even giants of technology like Lord Kelvin and James Clerk Maxwell brought it to the fore without it being widely applied. Eastman Kodak Company has had several engineers who have promoted the philosophy within as well as outside the company. Kodak's commitment to this topic resulted in Blanding's 1999 book [12], which is the source for most of the exact constraint material in *The Elements of Mechanical Design*, but the other citations were also used.

Few college textbooks discuss exact constraint, although the 2003 and later editions of David Ullman's *The Mechanical Design Process* do [13]. Engineering professors and practicing engineers alike lament this dearth of exact constraint material in mechanical engineering education, although some specialized texts do cover it [10,11].

References:
7   Michael French, *Conceptual Design for Engineers,* Third Edition, Springer-Verlag, 1999: pp173.

8   Lawrence Kamm, *Designing Cost-Efficient Mechanisms,* Society of Automotive Engineers, 1993. Also published by McGraw-Hill, 1990.

9   Douglass L. Blanding, *Principles of Exact Constraint Mechanical Design, Second Edition,* Eastman Kodak Company, 1995.

10 Daniel E. Whitney, *Mechanical Assemblies: Their Design, Manufacture, and Role in Product Development,* Oxford University Press, 2004.

11 Alexander H. Slocum, *Precision Machine Design,* Prentice-Hall, 1992.

12 Douglass L. Blanding, *Exact Constraint: Machine Design Using Kinematic Principles,* New York: ASME Press, 1999.

13 David G. Ullman, *The Mechanical Design Process, Third Edition,* McGraw-Hill, 2003.

14 Alan R. Parkinson and Alisha Hammond, "On the Robustness of Exactly Constrained Mechanical Assemblies." *Proceedings of the Design Engineering Technical Conferences 2003,* New York: American Society of Mechanical Engineers, 2003.

15 Alan R. Parkinson, Eric Pearce, and Kenneth Chase, "On the Design of Nesting Forces for Exactly Constrained, Robust Mechanical Assemblies." *Proceedings of the Design Engineering Technical Conferences 2003,* New York: American Society of Mechanical Engineers, 2003.

4. Load path visualization and force flow line methods are found throughout the design literature. One of the earliest is Juvinall [16], but Ullman [17] added labeling for type of stress.

References:
16 Robert C. Juvinall, *Engineering Considerations of Stress, Strain, and Strength,* McGraw-Hill, 1967.

17 David G. Ullman, *The Mechanical Design Process, Third Edition,* McGraw-Hill, 2003.

18 M. J. French, "An annotated list of design principles." *Proceedings of the Institution of Mechanical Engineers,* Institution of Mechanical Engineers, Vol. 208 (1994): pp229-234.

19 Michael French, *Form, Structure and Mechanism,* Springer-Verlag, 1992.

20 Crispin Hales, *Managing Engineering Design,* Essex, England: Longman Group, 1993.

21 Kurt M. Marshek, *Design of Machine and Structural Parts,* John Wiley & Sons, 1987.

22 G. Pahl and W. Beitz, Engineering Design, A Systematic Approach, Second Edition, London: Springer-Verlag, 1996.

23 Stuart Pugh, *Total Design, Integrated Methods for Successful Product Engineering,* Addison-Wesley, 1990.

5. Use of the term triangulation is spotty in design circles, but the concept is not. The term is descriptive.

References:
24 Kurt M. Marshek, *Design of Machine and Structural Parts,* John Wiley & Sons, 1987.
25 Douglass L. Blanding, *Exact Constraint: Machine Design Using Kinematic Principles,* New York: ASME Press, 1999.

6. Designing for uniform stress is commonly taught for structures, perhaps less so in design.

References:
26 Kurt M. Marshek, *Design of Machine and Structural Parts,* John Wiley & Sons, 1987: pp47.

7. Self-help principles are classified and defined somewhat differently by the few authors who present them. I have eliminated the more abstract names and describe them by effect.

References:
27 G. Pahl and W. Beitz, *Engineering Design, A Systematic Approach, Second Edition,* London: Springer-Verlag, 1996.
28 Alexander H. Slocum, *Precision Machine Design,* Prentice-Hall, 1992.
29 Crispin Hales, *Managing Engineering Design,* Essex, England: Longman Group, 1993.

8. This discussion of friction is limited to first principles applicable to mechanical design. For example, lubricants are not discussed. This is not because they are not important. But lots of mechanical design uses no lubricants, and their use is often rather utilitarian. If you use hydrodynamic, hydrostatic, or boundary lubrication, you must understand these special phenomena as well.

Much of this section comes from recommendations by practicing engineers, but quite a lot is nicely summarized by French [33]: "Prefer pivots to slides and flexures to either."

References:
30 Sastry Ganti, "Predicting Lock-up in Slide Bearings," *Machine Design,* Cleveland, OH: Penton Media, January 10, 2002: pp 94.
31 Robert Leibensperger, "The Conquest of Friction." *Mechanical Engineering,* New York: American Society of Mechanical Engineers, November 2003: pp46-49.
32 Ferdinand P. Beer and E. Russell Johnston, Jr., *Vector Mechanics for Engineers, Statics and Dynamics,* McGraw-Hill, 1988: pp 342.
33 Michael French, *Conceptual Design for Engineers, Third Edition,* Springer-Verlag, 1999.

34 Dava Sobel, *Longitude: The True Story of a Lone Genius Who Solved the Greatest Scientific Problem of His Time*, New York: Penguin Group, 1996: pp104-109.
35 William J. H. Andrewes, *The Quest for Longitude*, Cambridge, MA: Collection of Historical Scientific Instruments, 1996

9. Three dimensional solid modeling software has been built primarily "to speed the creation of models and working drawings. These systems support the construction phase of the design process." [37] That is exactly the phase addressed in *The Elements of Mechanical Design*. Controversy continues over proper—and improper—use of design software, but all the mechanical designers I interviewed use 3D solid modeling for design.

References:
36 David G. Ullman, "The Importance of Drawing in the Mechanical Design Process." *Computer and Graphics*, Vol. 14, No. 2 (1990): pp263-274.
37 Ellen Yi-Luen Do, "The Right Tool at the Right Time." Ph.D. thesis, Georgia Institute of Technology, 1998.
38 Eugene S. Ferguson, *Engineering and the Mind's Eye*, Cambridge, MA: The MIT Press, 1992.

10. References:
39 Ben-Zion Sandler, *Creative Machine Design*, Paragon House, 1985.
40 Stuart Pugh, *Total Design, Integrated Methods for Successful Product Engineering*, Addison-Wesley, 1990.

12. References:
41 Kurt M. Marshek, *Design of Machine and Structural Parts*, John Wiley & Sons, 1987.
42 Joseph Edward Shigley and Charles R. Mischke, *Mechanical Engineering Design, Fifth Edition*, McGraw-Hill, 1989.

13. References:
43 Henry Petroski, *TO ENGINEER IS HUMAN, The Role of Failure in Successful Design*, New York: St. Martin's Press, 1985.
44 Eugene S. Ferguson, *Engineering and the Mind's Eye*, Cambridge, MA: The MIT Press, 1992.

16. References:
45 Erik Oberg, Robert E. Green, and Christopher J. McCauley. *Machinery's Handbook, 25th Edition*, Industrial Press, 1996.

17. References:
46 Chi-Teh Wang, *Applied Elasticity, New York:* McGraw-Hill, 1953.
47 M. J. French, "An Annotated List of Design Principles." *Proceedings of the Institution of Mechanical Engineers*, Institution of Mechanical Engineers, Vol. 208 (1994): pp229-234.

18. References:
48 Joseph Edward Shigley and Charles R. Mischke, *Mechanical Engineering Design, Fifth Edition*, McGraw-Hill, 1989.

21 and 22. References:
49 G. Boothroyd and P. Dewhurst, *Product Design for Assembly*, Wakefield, RI: Boothroyd Dewhurst, Inc., 1987.
50 Daniel E. Whitney, *Mechanical Assemblies: Their Design, Manufacture, and Role in Product Development*, Oxford University Press, 2004.
51 Geoffrey Boothroyd, Peter Dewhurst, and Winston A. Knight, *Product Design for Manufacture and Assembly*, CRC Press, 2002.

23. References:
52 Ian Stewart, *What Shape Is A Snowflake*, New York: W. H. Freeman and Company, 2001: pp 32.

Index